Herbs

OF BRITAIN AND EUROPE

Bob Press

Illustrated by Christine Hart-Davis

B L O O M S B U R Y

LONDON • OXFORD • NEW YORK • NEW DELHI • SYDNEY

Bloomsbury Natural History

An imprint of Bloomsbury Publishing Plc
50 Bedford Square   1385 Broadway
London   New York
WC1B 3DP   NY 10018
UK   USA

www.bloomsbury.com

First published by New Holland (UK) Ltd 2002
This edition first published by Bloomsbury 2016

British Library Cataloguing-in-Publication Data
A catalogue record for this book is available from the British Library.

Library of Congress Cataloguing-in-Publication data has been applied for.

ISBN: PB: 978-1-4729-1646-4
ePDF: 978-1-4729-4026-1
ePub: 978-1-4729-1647-1

2 4 6 8 10 9 7 5 3 1

Designed and typeset in UK by Susan McIntyre
Printed and bound in China by C&C Offset Printing Co., Ltd

To find out more about our authors and books visit www.bloomsbury.com.
Here you will find extracts, author interviews, details of forthcoming events
and the option to sign up for our newsletters.

Frontispiece: Rosemary, p.75

Contents

Introduction

This book is intended to introduce the reader to 150 of the most common, important and widely used herbs by describing the plants themselves, indicating which parts are used and for what purpose. It is not a herbal in the true sense, since it does not attempt to describe preparation methods, precise dosage or other information vital to the safe and successful use of herbs. Such information can be found elsewhere if required, although the reader is advised that a qualified practitioner should always be consulted before use. Despite the necessary warnings, herbs are a fascinating, useful and, above all, enjoyable area to investigate.

The relatively recent resurgence of interest in various alternatives to standard medical treatment, as well as the popularity of foreign cuisines, signals a willingness on behalf of the public to take a more active interest in herbs. Yet almost everyone already reaps the benefits of herbs and their products, whether knowingly or not. Herbs can be employed in a variety of ways, which include their use for medicinal, therapeutic and culinary purposes. In commercially produced medicines, in health and beauty products, in preserved and ready-prepared foods and in recipes prepared at home or in restaurants, herbs are an everyday part of our lives.

Medicinal herbs were originally regarded as the simple ingredients of medications, hence the old term *simples* for such preparations. Combinations of such simples formed the compound medicine that would be prescribed for sufferers. Once the only remedies available to the population at large, simples have long been a mainstay of domestic medicine. Although some have been shown to be without value or even dangerous, others have been proved effective and a surprising number have a long, unbroken history of use. As such they are as valuable today as they ever were.

Therapeutic herbs are those used in aromatherapy and other holistic preparations – they are a source of the various essential oils used in aromatherapy. Although the unique properties of essential oils were recognised some 5,000 years ago, it is due to the recent interest in alternative remedies that aromatherapy is becoming an increasingly popular method of promoting good health and wellbeing. Essential oils are extremely concentrated and are therefore used in very small amounts in massages, aromatic baths, inhalations, compresses and various cosmetic preparations.

The use of herbs for culinary purposes began for purely practical purposes – to improve the keeping qualities of meat or, where this failed, to disguise or

▲ Juniper, p.24

make more palatable the poor quality and often rotten food that was all to be had. Modern technology and health regulations such as sell-by dates have improved this state of affairs immeasurably, yet culinary herbs are still used as widely as ever. Many herbs make food more palatable by easing digestion. Indeed, they are regarded as essential ingredients in a vast array of dishes and are of great commercial importance as preservatives.

What are herbs?

The word 'herb' can be used in a variety of ways. In a scientific sense it refers to plants which lack woody tissues, but as an everyday term it also refers to plants used for culinary or medicinal purposes. This can cause confusion since many herbs in the latter sense are woody plants such as the shrubby Lavender or even trees like Willow. Even greater confusion arises over the distinction between herbs (in the everyday sense) and spices.

Herbs are plants, or parts of plants such as roots, leaves, flowers or fruits, which are used to prepare medicinal items, in therapeutic medicine, and to flavour foods. They are usually thought of as being fragrant but this is not always the case; many herbs have no particular fragrance whereas some are very strong smelling indeed. The word 'herb' also covers prepared products of herbs. Although a distinction between plant and product should strictly be made, it generally makes little difference. The dried and fragmented leaves of Oregano obviously come from the plant of the same name but in cases where the products have an appearance or a name very different from that of the plants from which they are derived, the two may not be easily linked. Mustard is derived from two different plants (*Brassica nigra* and *Sinapis alba*) while Neroli is obtained from the Seville Orange tree.

Spices are generally regarded as the hard parts of aromatic plants, usually of tropical origin. They include roots and stems (Ginger), bark (Cinnamon), dry fruits (Cardamom) and seeds (Pepper) but some spices are actually soft parts such as flower buds (Cloves). Equally, the hard parts of many temperate plants provide herbal products and can be regarded as spices, such as Cumin seeds. Another term that can include examples of herbs and spices is 'condiment', usually defined as a seasoning that gives relish to food. This is a general term which includes inorganic substances such as salt and in the last century even vegetables such as onions and turnips were counted as condiments. In the medicinal sense even less distinction is made between herbs and spices. In addition there are a number of herbal products that fit neither category well, such as gums, resins and essential oils, and all medicinal plants tend to be referred to simply as herbs.

Perhaps the clearest difference between herbs and spices is that herbs tend to be mild-tasting and are usually more effective as flavourings when used fresh, while spices tend to be stronger and more pungent and are usually used dried. However, there is so much overlap between herbs and spices that any precise distinction is rather meaningless and plants from both categories are included in this book.

Names of herbs

Plants, and therefore all herbs, often have two names: a popular or common name and a scientific name. Scientific names, which are the international currency of botanical nomenclature, are the more reliable and informative of the two and it is well worth becoming familiar with them and with their correct application.

A scientific name consists of two elements, showing the genus and the species to which the plant belongs. The name not only labels or identifies the plant but can provide information about the plant itself or about its relationships. Thus the Dog Rose, *Rosa canina*, and the Apothecary's Rose, *Rosa gallica* cv. *officinalis*, are separate species both belonging to the rose genus *Rosa* and therefore sharing many similar characteristics. The specific name *canina* means 'of dogs' and refers to the supposedly inferior uses of this plant compared to other roses. The name *gallica* means 'of France' and refers to the origin of that species. The cultivar name *officinalis*, which means a medicinal plant sold in shops, clearly indicates the general use to which this species was put.

Common names are more familiar and can also be informative but have various drawbacks. They vary in different languages and are not always applied consistently even within the same areas of a country. A plant may have several common names, or share the same common name with another, completely different plant. A member of the Saxifrage family, Biting Stonecrop (*Sedum acre*) is equally well known as Wall Pepper but is entirely unrelated to the Sweet Pepper which belongs to the Potato family. Cinnamon and Cassia are both members of the genus *Cinamomum* although the common names give no indication of this and the fruit of Thorn-apple (*Datura stramonium*), while spiny, is otherwise not at all applelike as the name suggests. With herbs there is the additional complication of the products – that is the parts we think of and actually use as herbs and spices – having completely different names from the herbs from which they are derived or when a herb yields more than one product. Paprika is obtained by grinding the fruits of the Sweet Pepper (*Capsicum annum*) and cayenne from its close relative the Chile Pepper (*Capsicum frutescens*). White pepper and black pepper are both obtained from the same plant (*Piper nigrum*). In this book, the entries are usually listed under the common name of the plant rather than the name of the product. This will enable you to more easily find the same plants in other reference works, such as guides to identification. A few entries are listed under the product name where this is better known in a herbal sense than the plant. In these cases both names are included in the text and can also be found in the index.

The safe use of herbs

There is a tendency nowadays to view herbalism and especially homeopathic medicine as providing 'safe treatments', vouchsafed by hundreds, even thousands, of years of use during which time any treatments with harmful effects have been eliminated. While this view may be correct to some extent, it does not mean that herbal treatments are without risk or that they can be used with abandon, and the warnings accompanying many of the species in this book should be heeded. The situation is further complicated by the fact that some individuals will, almost inevitably, be allergic to particular herbs and may suffer bad reactions to treatments. Before embarking on any treatments other than simple, everyday ones such as tonics, it is essential to consult a doctor or other qualified practitioner.

The individual entries in this book describe which parts of the herb are usually used and in what form. Dosages are not given as they can vary depending on a number of factors such as the precise condition being treated, the age and general health of the patient and even the quality and origin of the herbal material available. Again, the safest way to determine the correct dosage is to consult an expert.

Finally, a warning about identifying herbs is needed, especially when these are gathered in the wild. Obviously, it is vital to use the correct species in any herbal preparation or culinary flavouring. The required herb may have very similar relatives with which it is easily confused. In many cases these relatives are themselves innocuous, but some are harmful or even acutely poisonous. A good example is the Carrot family which contains many common herbs such as Sweet Cicely but also very poisonous plants such as Hemlock. The concise descriptions of the herbs in this book contain some, but by no means all, of the diagnostic features for each species. Unless you are totally familiar with a herb always check its identity in an authoritative botanical guide and, once again, seek expert advice.

Growing herbs

Herbs are extremely rewarding plants to grow. Not only can they look and smell wonderful but they are also very useful. A good place for cultivating a variety of herbs is a sunlit, sheltered area with well-drained soil, although some herbs are extremely resilient and will grow in almost any conditions.

A popular style of herb garden is a formal one. This, as its name suggests, is laid out in well defined patterns and geometric shapes, often outlined with stone or bricks. For those prepared to devote a great deal of time and attention to such a garden, the individual beds can be surrounded with dwarf hedges instead, for example of Box or Lavender.

Less formal plantings can be made by growing herbs together with flowers and vegetables in an ad hoc way. Many herbs are well-behaved plants that fit easily into borders and are just as attractive and useful grown in this way. It is a sensible idea to ensure that all of the herbs can be reached easily for regular harvesting.

▲ Red Bergamot, p.74

The combinations of herbs and variations of design are endless and it is simply a matter of personal choice and, of course, the space available. Careful planning will produce a satisfying balance of colour, texture and aroma. Even in a tiny plot a small bed can be used to grow a limited but varied selection of herbs by grouping small plants together. It is not necessary, however, to have a garden at all in which to grow herbs. They can be grown almost anywhere, from hanging baskets to kitchen windowsills. Even a large pot outside the door is sufficient for a few culinary herbs to supply the kitchen.

However and wherever they are grown, remember to balance the numbers of any herbs against the extent to which they will be used. A whole bed of Thyme or Chives may be needed whereas one or two plants of Parsley will suffice. Some species, such as Ground-elder and Mint can be invasive and are best grown in a separate bed, or within a large flower pot or bucket which has had the bottom knocked out before being sunk in the soil.

Artificial chemicals should never be sprayed or applied to herbs. Few suffer from any serious pests and they can be further protected by 'companion planting' – a system in which one plant benefits from being grown in proximity to another. Aromatic herbs are known to be especially effective in this role. Common pairings include Rosemary with Sage and Coriander with Chervil.

The precise selection of herbs grown in a garden will obviously depend on a variety of conditions such as the type and acidity of the soil but, whatever conditions prevail, not all herbs can be grown together. It is often stated that herbs in general need to be grown in full sun but obviously woodland species do best in shade and, in fact, many other herbs such as

Growing herbs

Chervil, Sweet Cicely and Valerian grow happily in partial or dappled shade. Southern herbs, especially those native to the Mediterranean region, do need plenty of sunshine, not only to grow well but to fully develop the aromatic oils which give their foliage its characteristic scent. They also prefer light, well-drained soils similar to those of their home region. They can be susceptible to frost in northern climes but with a little protection will grow successfully in many areas. Conversely, some northern species are unhappy in regions with long, dry summers and warm winters and may fail to establish. Spearmint, for example, often grows poorly in Mediterranean countries. In cooler, northern areas, frost-resistant or hardy perennial species are obviously most suitable but tender herbs can still be grown, either in cold frames or by being moved indoors during the winter.

Collecting, harvesting and storing herbs

The garden is the best place to collect herbs as the plants are conveniently to hand but many herbs can also be collected in the wild. Searching for herbs in the fields and hedgerows can be fun but if you do undertake such an expedition there are some additional rules to follow. Do not collect near main roads or in areas where pesticides or other chemicals may have been used. Damage the plants as little as possible by taking only the parts you require, leaves or flowers for example, and leave the plant to continue growing. Remember that in the U.K. the law forbids the uprooting of any plant without the express permission of the landowner and also forbids the collecting of any part (including seeds) of some rare species. Until you are sufficiently confident in your own ability to accurately identify the plants, do collect in the company of someone who can. If there is any doubt whatsoever about the identity of a plant, leave it alone.

Other than observing these precautions, the rules for gathering herbs are the same for both wild and cultivated plants. Choose strong, healthy-looking specimens. Always handle herbs carefully and avoid bruising them; they are best carried in an open container such as a basket, never in plastic bags. The ideal season depends to some extent on the species and the parts to be gathered and some herbs, especially medicinal ones, need special treatment to preserve and store them. However, the following general guidelines apply to most herbs, including the more commonly used culinary ones.

▲ Garlic Mustard, p.31

HARVESTING FOR IMMEDIATE USE

When using fresh herbs pick only the quantity needed and use immediately. Choose young, tender growth – a good method for perennial species is to pinch off the tips of the main shoots to encourage bushy growth from side-shoots for later use. Otherwise pick off lower leaves before they become too old to use.

HARVESTING FOR LATER USE

The best time to harvest for later use is on a sunny day after all the dew has dried out and before the sun is too strong to reduce any oil content in the plants. Again, only gather as much as can be processed quickly or the quality of the final products will be reduced. As a rule, aromatic plants are best harvested just before the flowers emerge, while herbs in which the leafy, flowering tops are used are best just after the flowers have opened. When just the flowerheads or individual flowers are being collected, pick when the flowers are well open but before they begin to fade. Seeds are collected as they ripen and it is often easiest to simply cut the whole head and handle it by the stalk. Roots are usually dug up in the autumn, after they have had time to build up reserves and are at their plumpest.

DRYING HERBS

The best place to dry herbs is in a dark, warm and airy place such as an attic or warm out-house. Good air circulation is essential to prevent them from rotting while sunlight bleaches or discolours leaves and flowers as well as reducing any oil content. The fresh herbs should be stored and any less than

13

perfect parts discarded. Wash them only if necessary, and then briefly, and shake dry before spreading thinly on open racks or freely suspending them in bunches. Quick drying is vital to prevent loss of flavour and colouring. Artificial heat may be needed for this, especially in periods of wet weather. A number of herbs can be successfully dried in a microwave oven – experiment with the settings to obtain the best results.

Dried herbs are ready to store when the parts are brittle and break with a snap, but when drying whole plants remember that all parts do not dry at the same rate; petals dry more quickly than leaves, stems take even longer, and roots longest of all. Once completely dry, any remaining foreign bodies or unwanted parts should be picked out. Leaves can be stripped from the stems if required but should not be broken or crumbled as this causes them to lose flavour more quickly. Seeds are best shaken out of the dry heads.

STORING HERBS

Dried herbs can be stored in any convenient container as long as it is airtight. Glass jars are ideal but must be kept in a dark place as sunlight quickly removes both colour and flavour. Opaque containers are a good alternative and even tightly sealed plastic bags can be used. Herbs stored in this way will last easily until the next season's crop is ready, at which time it is best to replace them.

A suitably modern method for storing herbs is to freeze them. In this case the herbs should be fresh, having been carefully washed and dried if necessary, and packed into plastic bags. The full flavour is retained but do not use frozen herbs for garnishing food as they go limp when thawed.

The history of herbs and herbalism

Exactly when herbs were first used by man is unknown, though it was certainly in very ancient times, long before written records began. Precisely how the methods of preparing and applying herbs were discovered is equally a mystery. Trial and error alone seem insufficient and extremely risky given the poisonous nature of many plants and the elaborate procedures involved in many herbal preparations. The potential of some herbs must have been first recognised via their culinary use but history suggests their medicinal properties were recognised just as early. Certainly herbs were indispensable for making poor or nearly spoiled food palatable by masking the taste but it should also be pointed

The history of herbs and herbalism

out that in early herbalism there was no strict division into medicinal and culinary herbs. Many plants can be used for both purposes and there were others besides, such as cosmetics, dye plants, colorants and strewing herbs which also, quite properly, came within the domain of the herbalist and apothecary.

In earliest times, herbal lore must have been passed down through generations as an exclusively oral tradition, and it is only from the time when written records appeared that we can begin to trace the history and uses of herbs with any certainty. The first records were manuscripts which were themselves often based on older knowledge. Few originals of these manuscripts have survived, but their existence can be inferred from references made to them in later works. The earliest are said to be Chinese and written about 5,000 years ago, though there are doubts about the date. Somewhat more recent records, beginning around 2800 BC are known from India, Egypt, and the Middle East and there are many references to herbs in the Bible. In Europe, probably the earliest and most influential writings were those of the Ancient Greeks such as Aristotle (384–322 BC) and his pupils who wrote discourses on the properties of plants and their use in medicine. Aristotle maintained a garden with more than 300 different species of herb.

One of the most famous of the ancient authorities on herbs was Dioscorides, a Greek residing in Rome during the first century AD and whose great work, *De materia medica*, listed over 500 different plants. Like other early writers, Dioscorides described the healing properties of plants and, although the descriptions of the plants themselves were often minimal, the illustrations were realistic and life-like.

The Romans spread what knowledge of herbalism they had throughout their sprawling empire. However, after its fall in c. AD 410 and the dawn of the Dark Ages in the fifth century, the flourishing herb and spice trade was sharply curtailed in northern Europe, although it did continue to some extent between southern Europe and the East. The dissemination and appreciation of herbal knowledge also suffered, though monasteries continued to be centres of learning right up to the fourteenth century. The monks maintained both libraries and herb gardens and made detailed records of herbs and their uses which greatly contributed to existing knowledge.

Some important advancements were made during the Dark Ages, introduced through the works of Arab and Persian scholars. A Persian physician, Avicenna (979–1037), discovered how to distil volatile oils from herbs and flowers thus paving the way for the use of herbs in therapeutic treatments. There was also, at this time, something of a culinary revolution, especially in

The history of herbs and herbalism

Britain following the invasion of the Normans in 1066. The Normans brought with them their methods of cooking with herbs and spices and the cooking of the period would have most resembled that of the Middle East today.

The Renaissance in the fourteenth century brought advances in medicine and public hygiene and these were accompanied by advances in herbalism. In the fifteenth century the invention of printing using movable type had led to greater availability of herbals as those known previously only as manuscripts were produced in book form. The word 'herbal' means literally 'herb book' and some of the earliest survive as incunabula, the term used for books printed before 1501.

The sixteenth century was something of a golden age for herbalism, with many famous practitioners such as John Gerard, who was apothecary to King James I. He maintained a large physic garden in London where he grew his own supplies of herbs and is also credited with writing the famous *Herball* of 1597, though actually it is only a translation of an earlier, unpublished Belgian work.

Other changes which did nothing to advance herbalism also occurred during this century and two in particular had far-reaching effects. Herbs already had a long-established link with astrology but now an extreme version of this arose known as astrological botany. This school of thought attempted to associate the different herbs with particular planets and astrological signs. Diseases were also claimed to be caused and controlled by astrological influence. This rather absurd idea conveniently allowed an almost infinite range of herbs to be safely recommended for almost any condition. Astrological botany became particularly influential in seventeenth century England, where the well-known herbalist Nicholas Culpeper (1616–1654) was a principal exponent of it.

The second major change during the sixteenth century was the introduction of the so-called Doctrine of Signatures, an idea initiated by Theophrastus Bombast von Hohenheim (1493–1541), a doctor and professor at Basle who was better known as Paracelsus. He believed that the physical appearance of a plant indicated its medicinal properties. A good example is St John's-Wort, a herb used for treating wounds. According to the Doctrine, the translucent dots on the leaves signified the porosity of the skin and therefore the plant's ability to heal cuts on the human body. The reddish juice expressed from the flowers represented blood which served to confirm the herb's value in treating wounds.

This superstitious side to herbalism was not confined only to followers of astrological botany or the Doctrine of Signatures. The gathering and preparation of herbs had long involved considerable superstition and often the use of specific paraphernalia. Ancient Greek herb-gatherers were

▲ Sweet Bay, p.35

strongly advised to dig up Paeony root only at night lest they be seen by a Woodpecker which, enraged by the activity, would peck out their eyes. Such scaremongering may well have started as a means of safeguarding the herb-gatherers' trade by frightening off the uninitiated but similar stories became entrenched and were taken very seriously. A well-known example concerns the Mandrake, the root of which was thought to be in the form of a man or woman, complete with arms, legs and the ability to move.

Towards the end of the seventeenth century herbalism began to decline as the disciplines of medicine and botany started to separate. Herbals were superseded by Pharmacopoeias, which dealt with the medicinal aspects of plants, and Floras, which provided descriptions of them. Decreasing production of home-made remedies and their increasing replacement with proprietary and synthetic products bought at chemists' counters was also instrumental in the decline.

In the twentieth century medicine and botany more or less completely parted company and allopathy – what we think of as modern pharmaceutically based medicine – gained predominance over the older, traditional practices. As late as the beginning of this century, the bulk of allopathic medicines were still derived directly from plants but today almost all are pure, synthetic chemicals. There are certain advantages to this, such as consistency of quality, but there are also disadvantages such as undesirable side-effects.

It was as a reaction against the apparent shortcomings of allopathy that the opposing discipline of homeopathy came into being. Modern homeopathy was devised in the early nineteenth century by a German physician, Samuel Hahnemann. It uses only natural drugs and is essentially

The history of herbs and herbalism

based on two principles, that like cures like, and (unlike allopathy) that a drug becomes more effective when diluted. So a drug which in very large doses causes the same symptoms in a healthy person as a disease does in someone who is ill can be used to treat that disease when administered in minute doses. As an example, a patient with a bad cold accompanied by red and watering eyes might be treated with minute doses of Onion since, in large doses, Onion produces the same symptoms in a healthy person. While supporters of homeopathy claim various successes, detractors counterclaim that these are due either to psychosomatic effects or to the replacement of unsuitable synthetic medicines with harmless but ineffective homeopathic ones. Whatever the truth of the matter, there is as yet no hard evidence that the homeopathic theory is correct.

Despite the rise of allopathy, herbalism has never completely died out, even in the developed countries. It survives everywhere in some form or other, and in many areas is still a major source of aid and comfort and will doubtless continue to be so. In the East, modern medicine and traditional herbalism have continued side by side as parallel and often complementary systems and in many under-developed and Third World countries herbal remedies are the only ones widely available. Even in countries such as Britain and the USA a revival of herb use is occurring on a broad front.

Practices such as aromatherapy while perhaps not in the strict tradition of herbalism, are nevertheless closely akin to it and share various of its aspects. Although essential oils were used thousands of years ago by ancient civilisations for perfume, flavouring, healing and embalming, it was a French chemist, Professor Rene-Maurice Gattefosse who originated the term 'aromatherapy'. He used essential oils to treat wounded soldiers in World War I and claimed that such oils, particularly Lavender, were extremely effective in detoxifying the body of harmful substances caused by infected wounds. Madame Maury, a pupil of Gattefosse, recognised the potential of using plant essences for skin care and developed the massage techniques now usually associated with aromatherapy. Aromatherapy today is still undergoing the process of development and research is continuing into the unique properties of essential oils and their potential for healing.

All of this is occurring in a climate of renewed public interest, as people seek a return to what they vaguely think of as a safer, more sympathetic and 'natural' approach to medicine and health in general. Yet, as some of the examples given here show, the mere fact that a particular herb was advocated by a herbalist some centuries in the past is no absolute guarantee of its efficacy.

From its very beginnings, herbalism has included treatments that were charms rather than remedies and which covered a whole range of eventualities. The leaves of Rosemary placed in a vessel of wine were said to bring good luck and a speedy sale of said wine, while for the strong of stomach there was a love potion involving Periwinkle and House-leek pounded together with earthworms. While many recipes may seem, at best, harmless and, at worst, likely to induce violent illness in the patient, they at least show an awareness of the value of prevention rather than cure, an ideal which we hold to strongly today.

Modern research has shown, and continues to show, that although some of the herbs used by our ancestors are actually harmful others do, indeed, contain compounds which have the precise effects ascribed to these plants. So while we may safely disregard the wilder assertions of the early herbalists and take others with a pinch of salt, it seems there is still much to be learned about the value of herbs.

Glossary

Alkaloid Complex organic compounds containing nitrogen and obtained from plants. Their function in nature is unknown though they may provide protection against grazing animals. Many are poisonous, e.g. strychnine, but a large number are used medicinally as drugs, e.g. morphine and quinine.
Analgesic A pain-relieving drug.
Annual A plant that germinates, flowers, sets seed and dies in one year.

▼ Meadowsweet, p.39

Glossary

Antiseptic A substance that counters infection by preventing the growth of bacteria.

Axil The angle between a leaf and the stem.

Biennial A plant that germinates and grows in the first year and flowers, sets seed and dies in the second year.

Bulb An underground storage organ made up of scale-like leaves, fleshy and swollen with food reserves.

Bulbil A very small bulb produced in the leaf axils or in place of some of the flowers and capable of forming a new plant when shed.

Burr A fruit with spines or hooked, bristly hairs, dispersed by catching on the fur of passing animals.

Calyx All the sepals of a flower.

Capsule Dry fruit splitting when ripe to release seeds.

Carcinogenic Capable of producing cancer.

Catkin Slender inflorescence of small flowers, usually crowded together and wind-pollinated.

Corolla All the petals of a flower.

Disc floret A very small, tubular flower with equal lobes; typical of members of the Daisy family.

Diuretic A drug that increases urine production.

Essential oil The volatile oil produced by aromatic plants and providing their characteristic scent and taste.

Inflorescence A group of flowers and their particular arrangement, e.g. a spike or an umbel.

Involucral bract One of the bracts surrounding a head of small flowers or florets; typical of the Daisy family.

Lanceolate Shaped like the blade of a spear, widest below the middle.

Linear Very narrow, with parallel sides.

Mucilage A substance which swells and becomes slimy in water.

Narcotic A drug producing drowsiness leading to sleep and eventually unconsciousness.

Opposite Having a pair of leaves at each joint of the stem.

Ovate Egg-shaped, widest below the middle.

Palmate With lobes or leaflets spreading from a single point.

Panicle A complex type of inflorescence, often large and always branched.

Perennial A plant that flowers and sets seed each year; herbaceous perennials die down to ground level in winter but grow again in spring from underground storage organs such as bulbs.

▲ Common Evening Primrose, p.52

Perianth All the sepals and petals of a flower, especially when these are indistinguishable from one another.

Photosensitivity A reaction caused when light-sensitive chemicals are exposed to the sun.

Pinnate Having two parallel rows of lobes or leaflets.

Ray floret A very small tubular flower with one side of the apex extended into a long, petal-like strap; typical of the Daisy family.

Rhizome A horizontal, underground stem, sometimes forming a storage organ.

Semi-parasite A parasitic plant that obtains some of its food from another plant (the host); such plants have green pigment and their own root system and are often capable of living completely independently.

Spur A projection formed by the sepals or petals of a flower.

Stamen Male organ of a flower.

Stolon A creeping stem, often rooting from the joints where they touch the ground.

Style Elongated part of a flower's ovary bearing a sticky area (the stigma) which receives pollen.

Tannins Water soluble chemical contained in some plants; they are bitter-tasting.

Taproot A stout, main root, often acting as a storage organ.

Trifoliate With three leaflets.

Umbel A branched inflorescence, the branches of equal length and all radiating from the same point, like the spokes of an upturned umbrella; typical of the Carrot family.

Vermifuge A substance used to drive out worms.

▲ Broom, p.42

The Herbs

Juniper
Juniperus communis

Small tree or shrub up to 600cm
high. Prickly green foliage of female
trees studded with green, berry-like
cones ripening blue-black with a
dull bloom in the second or third
year. Native throughout most of the
temperate northern hemisphere. The
berries yield an antiseptic and strongly
diuretic oil used to treat cystitis and, in
aromatherapy, for detoxification. Diluted,
the oil can be rubbed on to the temples
to reduce pain from rheumatism and
neuralgia but the oil should be used
internally with great care, and never
during pregnancy. The berries flavour
gin and sauces.

White Willow
Salix alba

Silvery-grey tree up to 25m with
upswept branches. Leaves narrow,
silvery-hairy eventually becoming dull
green above. Catkins appear with the
leaves, males and females on separate
trees. Grows beside rivers and streams
throughout most of Europe, western
and central Asia. The fresh or dried
bark has long been used to treat colds,
aches and as a general painkiller. Its
effectiveness was originally explained
as being due to the White Willow's
obvious ability to grow unhindered in
damp places. Well-known nowadays as
containing the basis of aspirin which is
now produced synthetically.

Hop
Humulus lupulus

Perennial climber with twining stems up to 600cm high. Opposite leaves are large, usually with 3–5 lobes and bristly with stiff hairs. Plants unisexual, males with branched clusters of flowers, females with papery cones. Native to north temperate regions and often cultivated. The hops are used to counter liver and digestive disorders and to make a mild sedative. A hop-filled pillow is said to cure insomnia. Hops are best known, however, for their use in flavouring beer and are grown on a large commercial scale for the brewing industry. Only the fruiting heads are used.

Slippery Elm
Ulmus rubra

Deciduous tree up to 30m high. Leaves ovate up to 170mm long with an asymmetric base, sharply and doubly toothed margins and a rough upper surface. Fruits yellowish-green, papery. Native to damp woods of eastern North America. White inner bark slimy, hence the name of the tree. Powdered, it makes a thick, mucilaginous tea used internally to ease sore throats, coughs and ulcers and externally for wounds and burns. Removal of the inner bark damages or even kills the tree and increased demand has led to the use of the much less effective outer bark.

Nettle
Urtica dioica

Coarse perennial up to 120cm high, covered with stinging hairs. Leaves ovate, pointed and toothed. Flowers greenish, small, forming axillary spikes. Plants either male or female. Found throughout the northern temperate regions. Nettles are used to treat a variety of conditions, from gout to dandruff. One painful herbal remedy is urtication which is the rubbing or flogging of the limbs with Nettles to relieve arthritis and rheumatism. Aerial parts rich in vitamins A and C, iron and other minerals and are added to soups and salads or made into nettle pudding or beer.

Sandalwood
Santalum album

Small evergreen tree up to about 10m high, with slender, drooping branches and smooth, grey-brown bark. Flowers dull yellow turning reddish-purple later, bell-shaped. The fruit is *c.* 10mm across, dark red to black. A semi-parasite on other plants, possibly native to Indonesia out cultivated throughout tropical Asia. The heartwood provides medicinal extracts used to treat bronchitis and cystitis. An oil is extracted from it for perfumes and cosmetics. Supposedly an aphrodisiac, Sandalwood oil is thick, with a characteristic scent, and is used in aromatherapy to relieve tension and anxiety.

Chickweed
Stellaria media

Creeping annual up to 30cm long.
Stems much-branched, up to 90cm
bearing pairs of oval, pointed leaves.
Flowers have deeply divided, white
petals slightly shorter than the sepals.
A fast-growing weed native to Europe
but spread by man and now found
throughout most regions of the world.
Chickweed can be made into an
ointment or poultice for inflamed skin,
ulcers and chilblains. The leaves contain
vitamin C and their main culinary use is
as an addition to salads or even boiled
as a vegetable.

Clove Pink
Dianthus caryophyllus

Perennial with woody base and stems
20–50cm high. Leaves bluish-green,
narrow, in pairs. Flowers 35–40mm
across, in clusters of 1–5. Petals rose-
pink, shortly frilled at the margins. Native
to southern Europe and North Africa
but widely grown elsewhere and one
of the earliest herbs cultivated in Britain.
The fragrant flowers have a spicy scent
and flavour similar to that of true Cloves.
After removal of the bitter, narrow white
claw at the base, the petals are used to
flavour drinks, syrup, vinegar and salads,
or candied to decorate cakes.

Common Sorrel
Rumex acetosa

Perennial up to 100cm, with acid-tasting foliage. Leaves distinctive, arrow-shaped, with backward-pointing basal lobes. Flowers small with 3 broad inner perianth-segments which become red and papery in fruit; the sexes are on separate plants. Widespread throughout the northern temperate regions and cultivated in gardens. Leaves high in vitamin C, and oxalic acid giving them a tangy, acid taste. Taken continuously and over long periods, it can cause the formation of small stones of calcium oxalate. A salad and pot herb from ancient times, it tastes metallic if cooked in an iron pan.

Caper
Capparis spinosa

Shrubby perennial with straggly, sometimes spiny branches up to 150cm long and circular to ovate, fleshy leaves. Flowers white or purple-tinged, 4-petalled, with a mass of long, purplish stamens in the centre. Native to parts of the tropics and subtropics, and to cliffs and rocky places in the Mediterranean. Capers contain capric acid and are used as a condiment and in sauces such as sauce tartare. Only the unopened flower buds are eaten and these must be pickled in wine-vinegar to bring out the characteristic flavour.

Monkshood
Aconitum napellus

Erect perennial up to 100cm, with paired, blackish taproots. Leaves palmately lobed, with the lobes themselves deeply cut. Flowers mauve or bluish with 5 petal-like sepals, the upper one forming a cowllike hood. Found across much of Europe and northern Asia as far east as the Himalayas. All parts of this plant are poisonous, especially the roots, extracts of which were used to tip arrows. Despite this, it is still occasionally used to treat pain such as neuralgia, and for coughs, though only when prescribed by a doctor.

Golden Seal
Hydrastis canadensis

Woodland perennial 15–30cm high, growing in colonies. Rhizome sends up erect shoots, each with 2 lobed leaves and a single flower. Flower has greenish white stamens but lacks petals. Native to North America where it is becoming rare due to over-collecting. A traditional Indian remedy, it is reportedly one of the most used herbal products in the United States. Tea made from the bright yellow root is used to treat inflammation of the mouth, throat and other mucous membranes, eye and ear infections, jaundice and other conditions.

Opium Poppy
Papaver somniferum

Erect bluish-green annual up to 100cm. Leaves pinnately lobed. The 4 white, pink or purple petals sometimes nave a dark basal patch. The large, globose capsule has holes around the rim, acting like a pepper-pot to release tiny seeds. Probably originating in the Mediterranean region, it is now widespread both as a cultivated plant and a weed. Raw opium, obtained from the milky sap of the unripe capsules, yields various medicinal drugs including morphine and codeine as well as the addictive drug heroin. The ripe seeds which contain no drugs are used in cooking.

Bloodroot
Sanguinaria canadensis

Early flowering woodland perennial 15–30cm high with blood-red roots. Flowers white, up to 50mm across, with 8–10 petals. Flowers usually appear before the leaves. Native to eastern and central North America. The fresh root was formerly used to treat coughs, asthma and lung ailments, especially by the American Indians, and it is now known to contain anticancer and antiseptic alkaloids. However it is poisonous except in minute doses. Its only current commercial use is as a constituent of toothpastes and mouthwashes for combating plaque.

Garlic Mustard
Alliaria petiolata

Erect biennial up to 120cm high. Leaves distinctive, heart-shaped, toothed at the margins and smelling of garlic when bruised. Flowers white, with 4 petals 4–6mm long, are followed by slender fruits 6–20mm long. Found throughout Europe, North Africa and western and central Asia. It is little used medicinally but is antiseptic and will soothe bites and stings. The leaves of this herb taste mildly of Garlic and can be used as a substitute in salads and sauces.

Horseradish
Armoracia rusticana

Robust perennial up to 125cm, with a stout tap-root. Leaves dark green and glossy, stalked, oblong to ovate with toothed margins. Flowering stems leafy, erect and branching. Flowers white and 4-petalled. Native to southern Europe and western Asia but cultivated and naturalised in many temperate areas. The root is very pungent and acrid due to the presence of mustard oil. Its stimulatory and antibiotic properties are useful for urinary infections, gout, rheumatism and circulatory problems. Grated and mixed with cream it yields the well-known sauce. Young leaves can be added to salads.

CABBAGES

Black Mustard
Brassica nigra

Slender annual up to 100cm. Leaves pinnately cut, bristly, the terminal lobe much larger than the others. Flowers yellow with 4 petals 7–9mm long. Fruits, slender-beaked and containing dark brown seeds, are pressed against the stem. Widespread throughout most temperate regions and commonly cultivated. It contains an antibiotic oil and is used for poultices and foot baths to stimulate the circulation but can cause blistering of the skin. The hot-tasting condiment is obtained from the ground seeds, mustard powder being a mixture of Black and White Mustard with Saffron and coloured with Turmeric.

White Mustard
Sinapis alba

Similar in overall appearance to Black Mustard. Flowers and fruits slightly larger but the principal difference is the fruits which have broad beaks and spread out from the stem. Seeds pale. Found in most of Europe and the Near East, introduced in many other areas. White Mustard is used in similar ways to Black Mustard. However, the white form is milder and was also used internally as a laxative and tonic. Whole seeds are added to pickles and it is the seedlings of White Mustard which are used with Cress in salads and sandwiches.

Water-cress
Rorippa nasturtium-aquaticum

Perennial 10–60cm high with creeping, rooting stems growing upwards to flower. Leaves glossy, pinnate, with rounded leaflets. Flowers white, small, 4-petalled. Fruits slender with visible seeds in 2 rows on each side. Grows in shallow, usually running water throughout most of Europe, North Africa and western Asia. It can be confused with unrelated, poisonous species and in some areas wild plants may harbour a parasite – the liver fluke. The stems and leaves have a high vitamin and mineral content, particularly iron, and are used mainly in soups and salads.

Common Scurvy-grass
Cochlearia officinalis

Biennial or perennial 5–50cm high with long-stalked, kidney-shaped basal leaves in a loose rosette. Stem leaves clasping and fleshy. Flowers white or occasionally lilac. Found around the coasts of north-west Europe and in the Alps. The name derives from the early use of this plant to prevent scurvy, caused by vitamin C deficiency, among sailors on long voyages. The leaves do indeed have a high vitamin C content and were eaten fresh. A tonic was also made in the form of Scurvy-grass ale; the modern version is an infusion.

LAURELS

Cinnamon
Cinnamomum zeylanicum

Small, evergreen tree up to 10m high.
Leaves paired, ovate to elliptical, deeply
veined and dark, shiny green. Flowers in
small yellow branched clusters followed
by dark purple berries. Native to Sri
Lanka and the south-west coast of India,
cultivated elsewhere in the east and in
the West Indies. The spice is provided by
the fragrant, dried inner bark of young
shoots and sold either in the form of
curled, papery 'quills' or ground into
powder. Sometimes used to treat colds
and stomach disorders. Commonly used
as a sweet spice in baking and in drinks.

Cassia
Cinnamomum aromaticum

Similar to its very close relative
Cinnamon though making a larger tree.
Native to China and Burma, cultivated
in many parts of the subtropics. Spice
is obtained from 3 parts of the Cassia
tree which is also known as Bastard
Cinnamon. The dried inner bark is very
similar to Cinnamon and is often used
as a substitute although the quills are
coarser both in texture and flavour.
Dried leaves are used mainly in Indian
cooking and the dried, unripe fruits –
sometimes sold as Chinese Cassia buds
– are used to flavour sweets and drinks.

Sweet Bay
Laurus nobilis

Bushy evergreen tree reaching 20m. Leaves, wavy-edged and dotted with numerous oil glands, give off a strong spicy scent when bruised. Flowers yellowish-green, 4-petalled, with males and females on separate trees. Native to dry areas of the Mediterranean but widely grown elsewhere, both as a pot herb and as a clipped shrub. Formerly used as a strewing herb, an aid for digestion and for treating baldness. A culinary flavouring since ancient times, the leaves – dried or fresh – are an essential part of bouquet garni. Leafy branches formed the victors' wreaths of classical times.

Sassafras
Sassafras albidum

Deciduous shrub or tree 3–30m high with oval to 3-lobed, fragrant leaves developing yellow or red hues in autumn. Flowers in small, yellowish clusters appearing before the leaves. Fruits dark blue. Native to dense woodlands of eastern North America. Roots, bark, leaves and flowers were used medicinally, mainly as teas; dried leaves also formed the principal flavouring of soups and stocks. Used by indigenous peoples, and a major export during America's early colonial period, it is now banned in the United States as carcinogenic.

Star-anise
Illicium verum

Small, white-barked evergreen shrub or small tree 400–500cm high with large leaves and small, many-petalled yellow flowers. 'Fruit' consists of 8 single-seeded pods radiating from a central point to form a star. Native to southern China where it is also cultivated, and to north-east Vietnam. Fruits are harvested unripe and dried. Infusions of the fruits or seeds have been used to treat various complaints, such as sore throats and sickness. The fruits are used to flavour dishes such as beef and fish. The oil, Oil of Anise, flavours drinks.

Ylang-Ylang
Cananga odorata

Evergreen tree up to 33m with smooth, ashy bark and large, wavy-edged leaves. Flowers large, drooping with 6 narrow petals *c.* 75mm long. Greenish at first, they turn yellow and have a jasmine-like scent. Native from tropical Asia to Australia and cultivated elsewhere in the Far East. The flowers provide a heavily scented oil with relaxing, antidepressant properties, which is used to treat shock and, in aromatherapy, for tension and high blood pressure. Due to the strength of the oil, excessive or long-term use can cause headaches and nausea.

Biting Stonecrop
Sedum acre

Tufted evergreen perennial 5–12cm high. Leaves 3–6mm long, fleshy and swollen, crowded on short, sterile shoots, though more widely spaced on flowering ones. Flowers bright yellow. Grows on dry, alkaline soils, often on walls, and is found throughout Europe, northern and western Asia, North Africa and North America. It is slightly poisonous and though once a medicinal herb, taken internally it can cause blisters and is now used mainly as a corn remover. However, the dried and ground leaves of this plant have a hot, peppery taste and are sometimes recommended as a seasoning.

Raspberry
Rubus idaeus

Perennial producing erect, woody, biennial stems up to 150cm high and armed with weak, straight prickles. Leaves pinnate with 5–7 leaflets which are densely white-hairy beneath. Flowers white, nodding and borne in small clusters. Native to cool regions of Europe, northern and central Asia but widely cultivated. Raspberry-leaf tea is a cure for various childhood chills and fevers but is most often recommended during the later stages of pregnancy as it tones the muscles in preparation for childbirth. The popular edible fruits are used to treat kidney problems and anaemia.

ROSES

Dog Rose
Rosa canina

Deciduous, often scrambling shrub up to 500cm, with stout, hooked prickles and pink or white flowers. Fruits 10–20mm long, may be globose, ovoid or elliptical. Native to Europe, North Africa and parts of Asia, and naturalised in North America. Petals can be used for perfume, the leaves to treat wounds or as a laxative. The fruits, called hips, contain more vitamin C than even citrus fruits, and those of the Dog Rose contain most of all. They are used to make rosehip syrup and the increasingly popular rose-hip tea.

Apothecary's Rose
Rosa gallica

Spreading, deciduous shrub up to 100cm high with prickles. Flowers crimson and fragrant. Fruits bright red and globose. Native to Europe from Belgium southwards. It is both a medicinal and culinary plant and is used for flavourings, perfumes, powders and oils. The petals were, and occasionally still are, used for strewing and are added to pot pourri. The oil distilled from the flowers is used in aromatherapy to reduce tension, emotional stress and post-natal depression, and to cure insomnia.

Meadowsweet
Filipendula ulmaria

Tall perennial reaching 200cm. Leaves pinnate with pairs of large, toothed leaflets interspersed with much smaller ones. Flowers creamy-white, numerous and crowded into an inflorescence up to 250mm long. They have a heavy, rather cloying scent. Found throughout most of the temperate northern hemisphere. An infusion made from the fresh flowers can be taken for all those conditions for which aspirin would normally be used. This is hardly surprising since Meadowsweet contains the chemicals which produce aspirin and, indeed, the drug is named from the old latin name for the plant, *Spiraea*.

Agrimony
Agrimonia eupatoria

Perennial with mostly basal leaves and a long flower spike up to 150cm high. Leaves pinnate with 2–3 pairs of small leaflets between each pair of large ones. Fruits crowned with hooked bristles. Native throughout Europe and extending into Asia Minor and North Africa. The green aerial parts of this plant contain a nigh proportion of tannins which make it valuable as a gargle and digestive tonic. Modern research also suggests that it can greatly increase blood coagulation, a property which would justify its long history as a wound treatment.

39

ROSES

Salad Burnet
Sanguisorba minor

Perennial with a basal rosette of pinnate leaves and a leafy flowering stem 10–90cm high. Flowers small, greenish and tightly packed into globose or ovoid flowerheads 10–30mm long. They have 4 sepals but lack petals. Usually found on chalky grassland in most of Europe, parts of the Middle East and North Africa; naturalised in North and South America. Principally used as an addition to salads, the leaves have a mild cucumber flavour. The plant often remains green through the winter months and was much grown in times when fresh salad vegetables were scarce in winter.

Wood Avens
Geum urbanum

Perennial 20–60cm high with pinnate basal leaves and deeply lobed stem leaves. Flowers bright yellow. Fruiting heads are burr-like containing about 70 narrow, hairy fruits, each tipped with a hooked spine. Native to most of Europe and western Asia. The root contains the same oil as Cloves and is similarly antiseptic. It has also been used as a substitute for quinine to counter fevers and, like Agrimony, contains tannins which act as a digestive tonic. On a less practical note, it was worn by the superstitious as a charm against evil spirits and venomous beasts.

Lady's-mantle
Alchemilla vulgaris

Perennial 5–45cm high with a basal rosette of leaves, their kidney-shaped to circular blades with shallow, toothed lobes. Flowers yellowish-green, tiny, and grouped into dense, branched clusters. Grows in Europe, northern Asia and eastern North America, usually on acid or neutral soils. Used to prevent internal and external bleeding, particularly menstrual disorders. The leaves are made into a tea or used as a poultice; the fresh juice is said to cure acne. Cheese made from the milk of cows grazing this herb has a distinctive flavour. The young leaves can be added to salads.

Hawthorn
Crataegus monogyna

Thorny, deciduous tree up to 18m with deeply lobed leaves 15–45mm long, masses of white flowers and dark or bright red fruits. Found throughout Europe and much of Asia. Flowers and fruits are usually used as an infusion. Hawthorn helps restore high or low blood pressure to more normal levels and improves problems brought on by ageing of the heart, arterial spasms and angina. Treatment is only effective over a period of several weeks but unlike some other drugs there is no risk of habituation. A liqueur can be made from the berries.

PEAS

Broom
Cystisus scoparius

Much-branched shrub reaching 200cm, the slender, whippy twigs green and ridged. Small, trifoliate or undivided leaves often fall very early. Flowers yellow, pea-like and numerous. Pods 25–40mm long, flattened and oblong, are hairy on the margins and black when ripe. Widespread in most of Europe. Broom contains the alkaloid sparteine which is used in cardiac treatment and obstetrics and is a strong diuretic. However, the drug's composition is variable and it should not be used except under medical supervision. The pickled flower buds were an Elizabethan culinary item.

Liquorice
Glycyrrhiza glabra

Rhizomatous perennial with erect stems up to 120cm high. Leaves pinnately divided, the leaflets sticky beneath. Flowers bluish-purple, small, pea-like, borne in axillary spikes. Native to southern Europe and western Asia, also cultivated. The roots contain glycyrrhizin, which is 50 times sweeter than sugar. When changed into an acid, glycyrrhizin is similar to human adrenal hormones and its effects are like those of cortisone. Liquorice is used to treat a wide variety of problems, including arthritis, inflammation, stomach ulcers, fevers, diphtheria and tetanus and to flavour other bitter-tasting medicines. It is also a popular sweet.

Fenugreek
Trigonella foenum-graecum

Annual 10–50cm high, the trifoliate leaves with toothed leaflets 20–50mm long. Flowers yellowish-white and pea-like, tinged violet at the base. Pods 80–140mm long, narrow, slightly curved. Probably native to southwestern Asia but cultivated and widely naturalised in central and southern Europe and elsewhere. Rich in vitamins and minerals, particularly calcium. Reputedly an aphrodisiac, the seeds also contain steroid-like chemicals and in Chinese medicine are used to treat impotence and menopausal problems. The seeds are added to curries and preserves and, grown like Cress, they add a mild curry flavour to salads.

Alexandrian Senna
Cassia angustifolia

Small shrub only 50cm high. Pinnate leaves nave all of the lanceolate leaflets in pairs. Flowers yellow, 5-petalled and borne in loose, erect spikes. Pods 25mm wide, flattened. Native to semi-desert regions of Somalia and Yemen but cultivated in Asia. The sennas are probably the best known laxatives and Alexandrian Senna has a powerful action. The active chemicals, called anthroquinone glycosides, are found in the leaves and, particularly, in the pods. Senna is usually mixed in a syrup with other herbs and spices such as Cinnamon, Ginger or Liquorice to make it more palatable.

Tamarind
Tamarindus indica

Densely foliaged evergreen tree up to 30m high. Leaves 5–10cm long, pinnately divided into 10–20 pairs of closely set leaflets. Pods long and pendulous, up to 200mm, containing seeds embedded in a yellow pulp. Possibly native to tropical Africa but unknown in the wild, it is widely cultivated in India and other tropical areas. The edible pulp surrounding the seeds is used as a gentle laxative. Used locally for fevers brought on by the hot winds. It is rich in glucosides and citric, tartric and malic acids. The sharp flavour makes it a useful addition to drinks and preserves.

Nasturtium
Tropaeolum majus

Sprawling or climbing annual with stems growing up to 200cm long. Leaves parasol-shaped. Stalk attached to the centre of the blade. Flowers orange, yellow or red, 25–60mm across with a backward-pointing spur. Fruit 3-lobed. Native to Peru but very widely cultivated, in various colour forms. All parts rich in vitamin C and antibiotic sulphur compounds which combat infection. The peppery-tasting flowers can be added to salads, the flowers and leaves used for tea, and the young pickled fruits as a substitute for Capers though they are purgative when eaten in excess.

Rose-scented Geranium
Pelargonium graveolens

Perennial up to 100cm high, with fragrant palmately lobed leaves. Flowers mauve or pink, carried in loose umbels. Native to South Africa but cultivated in most parts of the world. One of the ornamental garden Geraniums, and not to be confused with the genus Geranium containing the Crane's-bills. The essential oil is used in aromatherapy massages for relieving premenstrual tension and fluid retention. This species has rose-scented foliage and the fresh or dried leaves add flavour to jellies, cakes and puddings. Other species yield a range of fragrances such as lemon and mint.

Castor-oil-plant
Ricinus communis

Robust annual or spreading, shrub-like plant up to 400cm, depending on the climate. Leaves palmate, up to 600mm across with 5–9 lobes. Spikes stout, containing greenish male flowers below prickly-looking, reddish clusters of females. Fruits 10–20mm, globular and spiky. Native to the tropics out widely grown and naturalised in many areas. Castor oil is obtained from the crushed seeds and is used in engine fuels, lubricants, paints, varnishes and insect repellants. It is also a mild laxative. The seeds contain the poison ricin and are extremely toxic.

Rue
Ruta graveolens

Aromatic evergreen shrub up to 45cm high, with grey-green pinnately divided leaves. Flowers 20mm across, each of the 4 yellow petals has an incurved, hooded tip. Widely grown and sometimes naturalised outside its native eastern Mediterranean region. The aerial parts yield an oil used in small doses to strengthen weak blood vessels, promote menstruation and treat colic. An ointment is used for sprains and bruises but the sap can cause a strong photosensitive reaction on contact with the skin. The plant is sometimes taken as a bitter tea.

Neroli
Citrus aurantiacum

Small evergreen tree up to 10m high, with broadly elliptical, thin but leathery leaves 7.5–10cm long. Flowers white and fragrant with 4–8 petals and numerous stamens. Fruit 75mm diameter, orange, globular, thick-skinned and bitter-tasting even when ripe. Originating in the Far East but cultivated commercially in Europe and elsewhere. Distilled from the flowers of the Seville Orange, Neroli oil is heavy and very strongly scented. It is used in aromatherapy to sooth nervous tension, to encourage sleep, and is particularly recommended for rejuvenating the skin.

Frankincense
Boswellia carterii

Evergreen shrub or small, papery-barked tree up to 600cm. Leaves pinnate with spike-like clusters of small, waxy white flowers in their axils. Native to Somalia and Arabia. One of the gifts of the Magi and a major ingredient of sacred incense, Frankincense is extracted from the gum-like resin of several closely related trees of which this species is one of the most important. In aromatherapy, the oil is considered a rejuvenant, soothing and inducing a meditative mood to combat anxiety, respiratory problems and ageing skin. It is usually blended with other oils.

Surinam Quassia-wood
Quassia amara

Small tree reaching only 600cm high. Leaves pinnate, divided into 5 leaflets. Flowers red, tubular and borne in clusters at the tips of the twigs. Native to tropical America. Both the bark and the roots contain bitter principles which were used to treat dysentery, and which are the source of the mixer drink bitters. It should not be confused with a green-flowered tree from the West Indies, also called Quassia-wood, from which wood chips, boiled in water, provide an insecticide.

BUCKTHORNS / LIMES

Cascara Sagrada
Rhamnus purshiana

Small deciduous shrub or tree up to
12m high with pale greyish bark. Leaves
50–150mm long, prominently veined
with clusters of tiny greenish flowers in
the axils. Berries 8mm diameter, purplish-
black when ripe. Native to western North
America. The bark provides a gentle
laxative which does not require repeated
doses. Combined with pleasantly
aromatic herbs, it is considered suitable
for frail or convalescent people and is also
used by vets for treating dogs. Originally
used by American Indians in California,
it is now exported to Europe where it
has replaced treatments derived from
local species.

Lime
Tilia × vulgaris

Tall, narrow-crowned deciduous tree
up to 46m high. Leaves 60–100mm
across, broad and heart-shaped, often
sticky with sap. Flowers yellowish-
white and fragrant hang in a cluster
beneath an oblong, wing-like bract. A
naturally occurring hybrid between two
European species and widely planted as
a street-tree. Tea made from the fragrant
flowers is recommended for nervous
disorders, migraines and insomnia.
Lime flower tea is also used to treat
colds and bronchial complaints. The
inner bark is used for kidney ailments
and coronary disease.

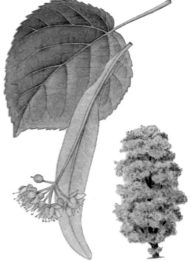

Marsh-mallow
Althaea officinalis

Densely grey-hairy perennial up to 200cm high, with large, toothed and sometimes palmately lobed leaves. Flowers, lilac-pink with shallowly notched petals 15–20mm long, form a tall spike. Native to Europe, North Africa and western Asia, introduced to North America. Marsh-mallow has a high mucilage content. The herb is used to reduce inflammation of the stomach. It also makes a gargle for throat and mouth infections. Sucking a Marsh-mallow stick is an old remedy for teething children and a sweet is made from the root.

Perforate St John's-wort
Hypericum perforatum

Rnizomatous perennial up to 100cm high, the woody-based stems with 2 raised ridges. Leaves up to 3cm, stalkless and dotted with numerous translucent glands. Flowers *c.* 20mm across, yellow, with numerous stamens. Widespread in temperate regions. The foliage has an antibacterial effect and is used as a dressing for deep wounds. The red oil, extracted by steeping flowers and leaves in vegetable oil, can be used externally for neuralgia, sciatica and burns but can cause a skin reaction. A major use of the herb is in the treatment of severe depression by inducing euphoria.

Sweet Violet
Viola odorata

Creeping perennial up to 15cm high
with a rosette of kidney-shaped leaves
and long, rooting stolons. Long-stalked
flowers are spurred and either dark
violet or white. Native throughout
much of Europe. The plant contains
methyl salicylate, from which aspirin is
derived and is a traditional treatment
for migraines and headaches. A syrup
made from the flowers, or tea from the
leaves, soothes coughs and bronchial
complaints. The flowers form the base
for perfume and for a wine said to be
good for hangovers; candied, they are
used as decorative confections.

Eucalyptus
Eucalyptus globulus

Large, fast-growing evergreen tree
up to 40m high, the grey-brown bark
peeling away in long strips. Juvenile
leaves bluish, opposite and clasp
the stem. Adult leaves dark green,
alternate and drooping. Flowers are
woody cups with numerous stamens
but no petals or sepals. Native to
Australia the Tasmanian Blue Gum
is also planted in many parts of the
world. Eucalyptus oil, distilled from
the adult leaves, is a strong antiseptic.
It is used in aromatherpy in inhalants
and decongestants, as a chest rub for
coughs, and as a common component
of cough sweets and pastilles.

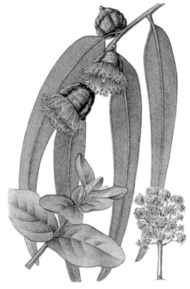

Allspice
Pimenta dioica

Evergreen tree reaching 9m in height, with opposite, leathery leaves 75–150mm long. Flowers 9mm across; cream and white, 4-petalled. Males and females are carried in clusters on different trees. Fruits ripen from green to dark purple. Native to the West Indies, Central and South America. So named because its flavour combines those of Cinnamon, Clove and Nutmeg. It is used in pickles, preserves and in baking, and is added to many spice mixtures. The berries are picked unripe and dried. The rind is the most aromatic part and the berries are ground immediately before use.

Cloves
Syzigium aromaticum

Small evergreen tree 10-15m high with paired leaves and clusters of bright red flowers. These are rarely seen on cultivated trees as it is the unopened flower buds which, when dried, form the cloves. Native to the Molucca Islands of Indonesia but also cultivated in Zanzibar, Madagascar and the West Indies and always grown near the sea. Oil of Cloves is an analgesic used, for example, to counter toothache. It is also used in the perfume industry. The flower buds are used both whole and powdered to spice a wide variety of culinary dishes.

Common Evening Primrose
Oenothera biennis

Biennial with a fleshy root and leafy stems up to 150cm high. Flowers 45–60mm, yellow, 4-petalled, fragrant, and open in the evening. Fruits long, slender containing numerous tiny seeds. Native to North America but naturalised in most of Europe. The leaves, seeds and root are used, either as an infusion or for extracting an oil. The effectiveness of this herb has been given credence by recent clinical testing. The oil is successful in treating a variety of conditions including premenstrual syndrome and hyperactivity, alcoholism, high blood pressure, arthritis and multiple sclerosis.

Witch Hazel
Hammamelis virginiana

Shrub or small, deciduous tree only 500cm high, with smooth greyish bark. Downy-hairy leaves are widest above the middle, with scalloped margins. Flowers yellow, appearing in clusters after the leaves have fallen. Petals *c.* 25mm long, strap-shaped and very narrow. Native to North America, often cultivated in gardens elsewhere. Astringent, tannin-rich tea made from the bark or twigs was used both internally and externally to maintain muscle-tone and to treat dysentery, cholera and other ailments. Modern, commercially distilled extracts and ointments are mainly used for minor bruises and scratches.

Nutmeg and Mace
Myristica fragrans

Evergreen tree reaching 40m high, with aromatic leaves and clusters of small, pale yellow flowers; males and females on separate trees. Fruit large and fleshy containing a kernel, the nutmeg, enclosed in a red net, the mace. Native to the Molucca Islands but cultivated in wet, seaside areas elsewhere, including the East and West Indies. Nutmeg and mace are dried and sold separately. Nutmeg is a fragrant, sweet spice used grated in cooking and as a digestive tonic, although it is toxic if taken in excess. Mace is similar but stronger and more pungent smelling.

Chinese Ginseng
Panax ginseng

Perennial up to 60cm, with a large root and a single whorl of palmate leaves at the top of the unbranched stem from which emerges a long-stalked umbel of yellowish-green flowers. A forest plant of north-east China. Most highly prized of the ginsengs, it is regarded as a unique tonic. The root contains saponins and steroids, hormone-like chemicals giving an 'adaptogenic' effect, which means returning the body to normal. Used to counter the weakening effects of age, stress or disease. Also thought to improve endurance and the ability to concentrate.

CARROTS

Chervil
Anthriscus cerefolium

Rather wiry annual up to 70cm high, with bright green, much-divided, pinnate leaves, the lobes deeply cut. Flowers small, white, in umbels. Fruits 7–10mm long including a slender beak of *c.* 4mm; narrow and hairless in cultivated plants. Probably native to south-east Europe and western Asia but widely cultivated and naturalised elsewhere including North America. The leaves of this culinary herb are used fresh. It is a principal ingredient of fines herbes mixtures but can be used alone for its delicate, slightly aniseed flavour.

Sweet Cicely
Myrrhis odorata

Softly hairy perennial reaching 200cm, with hollow stems and foliage smelling strongly of aniseed when crushed. Leaves 2–3-times pinnate with oblong-anceolate, toothed and white-blotched lobes. Fruits 15–25mm long, narrowly oblong and sharply ridged. A mountain plant from the Alps, Pyrenees, Apennines and Balkan mountains; cultivated and widely naturalised elsewhere. The leaves and aromatic seeds have an aniseed flavour and can be added to salads and whipped cream or eaten on their own. The chopped leaves can be cooked with tart fruit preserves and are a natural sweetener suitable for diabetics.

Coriander
Coriandrum sativum

Annual with solid, ridged stems 15–70cm high and 1–3-times pinnate leaves. Fruits 2–6mm, red brown and hard, the two halves not separating easily. Native to North Africa and southwest Asia but widespread elsewhere as both a crop and a weed. An ancient herb, known from at least 1500 BC. The leaves are sometimes used as a garnish or in curries but it is mostly the ripe seeds that are used for culinary purposes. They are an ingredient of many dishes, both sweet and savoury, and add spice to salads.

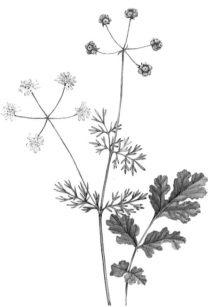

Anise
Pimpinella anisum

Strongly aromatic annual 10–50cm high. Lowest leaves kidney-shaped, middle leaves pinnate with broad lobes, upper leaves 2–3 times pinnate with narrow lobes. Flowers white giving way to ovoid or oblong, finely ridged fruits 3–5mm long. Native to the eastern Mediterranean and western Asia but widely cultivated. Anise yields an essential oil containing the compound anethole and is used commercially in toothpastes and cosmetics. The distinctively flavoured seeds are added to dishes such as curries and make a relaxing and very mildly narcotic tea which relieves tight coughs.

CARROTS

Ground-elder
Aegopodium podagraria

Stout and hairless perennial spreading by creeping rhizomes. Stems 40–100cm high, hollow bearing leaves divided into 3 leaflets; each leaflet may itself be divided into 3. Native to Europe and Asia, but introduced as a pot herb in some areas, including Britain. A traditional herb for treating arthritis, rheumatism and gout, to which the alternative name Goutweed alludes. An infusion or a poultice can be prepared from the leaves; an essence derived from the same parts is used in homeopathy. The young leafy shoots gathered just before flowering are edible.

Fennel
Foeniculum vulgare

Tall perennial up to 250cm, with a solid, polished stem. Leaves feathery, finely divided into numerous slender segments. Plant bluish-green except for yellow flowers. Native to the Mediterranean and southern Europe as far north as Britain but naturalised in many other countries. An ingredient of many proprietary cough medicines, an infusion of the seeds, or root, is taken for various minor ailments such as coughs, colic and lack of appetite. Diluted and unsweetened, it also makes a drink suitable for babies. The swollen bases of the leaf stalks can be served as a vegetable.

Dill
Anethum graveolens

Slightly bluish, strong-smelling annual 20–50cm high, resembling Fennel in its much divided, feathery leaves and yellow flowers. Fruits 5–6mm long, dark brown, and bordered with a pale wing; elliptical and flattened. Probably native to south-western Asia but widely cultivated and naturalised in many temperate regions. Oil of Dill is a mild sedative and dill water, made by infusing bruised seeds. It eases colic and is given to babies. An ancient culinary herb, Dill leaves are addea to fish dishes and cream sauces. The stronger-tasting, aromatic seeds are used to spice vegetables, pickles and vinegar.

Cumin
Cuminum cyminum

Slender annual 10–50cm high, the leaves divided into threadlike lobes. Flowers white or pink, 3–5 in each of the small, simple umbels which form the compound umbel. Fruits 4–5mm long, finely ridged. Native to North Africa and south-western Asia, cultivated elsewhere. Whole or ground Cumin seeds are a common ingredient of both Asian and North African cuisines and are frequently added to spice mixtures. It is commonly used in curries and is often added to chicken, lamb and beef. A traditional Indian drink includes Cumin seeds and Tamarind water.

Celery
Apium graveolens

Biennial up to 100cm high. Lower leaves pinnate, the upper divided into three leaflets. Flowers greenish-white, followed by ovoid fruits *c.* 1.5mm long. Found in Europe, Asia and North Africa. The cultivated Garden Celery (var. *dulce*) is less pungent-tasting than the wild plants and is most commonly used. The green leaves are reputed to lower blood pressure and can be added to salads while the seeds bring an aromatic flavour to stews and are often used in the form of celery salt. The swollen, blanched leaf stalks of this garden form are the well-known vegetable.

Parsley
Petroselinum crispum

Stout-rooted biennial 30–75cm high with a solid stem and sharply ascending branches. Leaves 3-times pinnate, shiny green. Flowers yellow. Probably native to south-east Europe or Western Asia but cultivated and naturalised in all temperate regions. Parsley is strongly diuretic, increases lactation and tones the uterus. It must not be used during pregnancy. The curly, frilled eaves of some cultivated forms are a popular culinary garnish. More than a mere decoration, they are a breath-freshener recommended against the lingering smell of garlic and are an excellent source of both vitamin C and iron.

Caraway
Carum caroi

Much-branched biennial 25–60cm high with hollow, faintly grooved stems. Leaves 2–3-times pinnate with segments divided into narrow lobes. Flowers white or pink. Fruits 3–4mm long, ellipsoid and ridged. Found in temperate regions of the Old World, often introduced. An ancient herb, Caraway is cultivated on a large scale today. Various proprietary medicines for colic, flatulence and lack of appetite contain Caraway. As well as adding flavour, the seeds are one of the best natural stimulants for appetite, and are widely used in cooking and baking.

Angelica
Angelica archangelica

Perennial up to 200cm high, with stout, hollow, green stems. Flowers greenish. Fruits have broad, corky wings. Found from northern Europe and Greenland to Central Asia; introduced in North America. A tea made from seeds or dried root combats anaemia, bronchitis, asthma and – by inducing a strong dislike of alcohol – alcoholism. The fresh leaves can be added to soups, fish and stewed fruit, the seeds to flavour liqueurs such as Chartreuse. Best known for the candied young stems of confectionery. Angelica has a high sugar content and diabetics are recommended to avoid it.

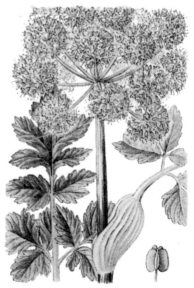

CARROTS

Lovage
Levisticum officinale

Stout, strong-smelling perennial
100–250cm high with 2–3-times pinnate
leaves up to 700mm × 650mm, the
lobes deeply and irregularly toothed.
Flowers greenish-yellow. Fruits 5–7mm
long, ridged and winged. Native to Iran,
cultivated and naturalised in Europe
and elsewhere. The dried leaves make a
tea for fevers but it should not be taken
during pregnancy or if suffering from
kidney disease. Plant has a rather strong
flavour resembling that of Celery. The
leaves and seeds can be added to many
dishes, especially vegetarian ones, and
the seeds are also used in breads and
other savouries.

Asafoetida
Ferula assa-foetida

Yellowish, foul-smelling perennial up
to 200cm high with stout, grooved
stems and thick roots. Leaves up to
350mm × 250mm are deeply 2–4-times
divided. Flowers yellowish followed
by fruits 12mm × 7mm. Native to Iran.
Asafoetida is a waxy gum-resin derived
from the root and stem of this and two
very similar species. It is also used in
veterinary medicine. As a condiment it
is mainly used in Persian, Afghan and
Indian cuisine, especially in vegetarian
dishes. The whole plant smells strongly
due to the presence of sulphur
compounds but the smell disappears
after boiling.

Wintergreen
Gaultheria procumbens

Creeping, evergreen shrub only 15cm high with dark, shiny leaves which are thick and leathery. Flowers white, waxy, drooping and bell-shaped. Berries bright red, persisting throughout the winter. Native to northern and eastern North America. The fragrant leaves contain methyl salicylate, which is similar to aspirin and is used to treat rheumatism. It is extracted in the form of an oil and the oil content of frosted leaves (which turn purple) is thought to be higher than that of unfrosted leaves. Natural oil of wintergreen has been largely replaced by synthetic compounds.

Cowslip
Primula veris

Glandular-hairy perennial with a basal rosette of finely hairy leaves 5–15cm long. Flowers, deep yellow with orange spots at the base of the corolla-lobes, hang in a cluster at the tip of a stalk 100mm–300mm high. Native to Europe and temperate Asia, mostly on lime-rich soils. Cowslip has sedative and expectorant effects and the roots are used to treat whooping cough, bronchitis ana arthritis. An ointment of the flowers treats spots and sunburn although some people may develop an allergic reaction known as primula dermatitis. The flowers are also used medicinally in various drinks.

OLIVES / GENTIANS

Jasmine
Jasminum officinale

Deciduous or semi-evergreen woody climber up to 10m. Opposite leaves pinnate with 5–7 leaflets, each 10–60mm long. Flowers usually white, sometimes flushed purple on the outside; fragrant and tubular. Native to south-west Asia but widely cultivated. Jasmine oil is an expensive but powerful fragrance obtained from the flowers. It is used by perfumers and is employed by aromatherapists to treat depression and respiratory problems. The oil can also be rubbed in to ease neuralgia. Jasmine tea is a mild sedative and a remedy for headaches.

Great Yellow Gentian
Gentiana lutea

Stout erect perennial up to 50–120cm, the large, ribbed leaves opposite and clasping; those towards the base of the stem form a rosette. Flowers yellow with 5–9 corolla-lobes. Confined to the mountains of central and southern Europe. All gentians contain extremely bitter principles; in this species they are obtained from the dried root. Once regarded as something of a universal panacea, the bitters are effective against a wide variety of digestive ailments and stimulate production of red blood cells. It should not be taken during pregnancy or if suffering from high blood pressure.

Bogbean
Menyanthes trifoliata

Aquatic plant of 12–35cm with leaves divided into groups of 3 leaflets and held above the water surface. Flowers, pink and white with fringed petals, are borne in spikes. Native to still water and bogs throughout most of the temperate northern hemisphere. A traditional tonic and purging herb, the leaves ana rhizome contain bitter compounds similar to those found in gentians. Recommended for a variety of complaints, including anorexia, it stimulates the appetite but can cause vomiting in large doses. Also used for flavouring ales and other alcoholic drinks.

Woodruff
Galium oderatum

Slender, fragrant perennial up to 45cm with creeping rhizomes and erect stems. Leaves usually 6–9 in each whorl, their margins with tiny, forward-pointing teeth. Flowers white and fragrant, each 4-lobed and 6mm across, form dense clusters. Found throughout much of Europe, North Africa and northern Asia. This traditional strewing herb contains coumarin, the compound which gives new-mown hay its distinctive scent. It is contained in some medicines to treat haemorrhoids and prevent thrombosis. The leaves and flowers make an excellent tea and are used to flavour wines for May-cups.

Lady's-bedstraw
Galium verum

Perennial up to 120cm high, with creeping stolons and much branched, 4-angled stems. Leaves very narrow, 8–12 in each whorl. Flowers 2–4mm across, bright yellow, 4-lobed forming a branched spike. Native to most of Europe and western Asia. Like its relatives Woodruff and Cleavers, Lady's-bedstraw was respectively a strewing herb and a wound herb but most of the old uses of this plant were connected with curdling milk. However, it is said to be less effective with modern milk, although it can still be used to impart a rich yellow colour to cheese.

Cleavers
Galium aparine

Rough and minutely prickly annual which scrambles through surrounding vegetation. Stems reach 180cm and are 4-angled with 6–9 leaves in each whorl. Flowers tiny, whitish and 4-lobed. Fruits 4–6mm across, burr-like, consisting of 2 fused globes covered with hooked bristles. Native throughout Europe, northern and western Asia. The dried aerial parts make an infusion used as a diuretic for cleansing the lymphatic system, reducing swollen glands and, traditionally, for cancer. It similarly helps against eczema and psoriasis and is a wash for sores and wounds. It is also recommended for treating dandruff.

Yellow-bark
Cinchona calisaya

Evergreen tree up to 12m high, with oval to oblong leaves. Flowers small, fragrant, pink and borne in clusters. Native to the eastern Andes; introduced to Asia where it is grown in plantations. The bark contains a large number of alkaloids, particularly quinine, which is used for fevers and heart problems and which is the most effective treatment for malaria. The alkaloids can be toxic and as a drug quinine should only be taken under medical supervision. It is also the flavouring used in tonic water.

Lungwort
Pulmonaria officinalis

Hairy perennial 20–30cm high with clumps of spotted, long-stalked leaves which enlarge after flowering, the blades reaching 160mm long. Flowers pink and blue, funnel-shaped and carried in terminal clusters on leafy stems. Found mainly in central and southern Europe, northwards to Britain and Sweden. Due to their supposed resemblance to lungs, an infusion of the white-spotted leaves was regarded as a remedy for pulmonary complaints. This traditional use has since been confirmed, the leaves containing soothing mucilage and silica which restore elasticity to the lungs.

BORAGES

Comfrey
Symphytum officinale

Erect, bristly perennial 50–120cm high.
Leaves are 150–250mm long, the upper
ones extending down the stem as
wings. Flowers 12–18mm long, purple-
violet, pinkish or white, tubular to
bell-shaped, are carried in coiled sprays.
Occurs in most of Europe. Comfrey has a
reputation as a healing herb, a poultice
of leaves or roots being effective for
bruises, ulcers and burns. It contains
allantoin which is absorbed through
the skin and stimulates repair of tissue
and bone. However, there is conflicting
evidence about potential carcinogenic
effects and this herb should not be
taken internally.

Borage
Borago officinalis

Erect, bristly annual 15–70cm high
with stalked basal leaves. Upper leaves
stalkless and clasping the stem. Flowers
200–250mm across, blue, with 5
spreading, pointed corolla-lobes. The
black stamens form a central cone.
Native to southern Europe but widely
cultivated and naturalised elsewhere.
The herb has a long history of medicinal
use, particularly as a tea for coughs and
in treating depression. It contains the
same active compound as Evening-
primrose and is thought to work by
stimulating the adrenal glands. The
young leaves are added to salads, the
leaves and flowers to summer drinks.

Vervain
Verbena officinalis

Erect perennial 30–60cm with tough, slender, 4-angled stems and opposite, pinnately lobed leaves 40–60mm long. Flowers 4mm across, pale pink, slightly 2-lipped and carried in long, slender spikes. Grows from Europe and North Africa to the Himalayas, and has been introduced to North America. It is a herb with a long tradition of magical and medicinal use. A tea made from the aerial parts is used to treat nervous exhaustion, headaches and migraine. It is also said to be effective against liver and gall-bladder disorders. It should not be taken during pregnancy.

Lemon Verbena
Lippia triphylla

Deciduous shrub up to 8m high in the tropics but much less in cooler regions. Leaves narrow, yellowish-green, borne in whorls of 3. Flowers pale lavender, 2-lipped and grouped in slender, terminal spikes. Native to South America but widely cultivated in the Mediterranean and other parts of the Old World. The aerial parts are strongly lemon-scented when crushed and are used as a tea to improve digestion. It is also said to counter depression, lethargy, migraine and vertigo. The essential oil extracted from the plant is used in perfumes and liqueurs.

Basil
Ocimum basilicum

Perennial or, in cool regions, an annual, reaching 50cm. Opposite leaves are hairless and slightly fleshy. Flowers white or mauve-tinged, 2-lipped and borne in whorls. The whorls form a loose spike. Several varieties, including a purple-leaved form, are cultivated. Native to India but cultivated in many parts of the world. The fresh eaves are the principal ingredient of pesto sauce. As well as being served with tomatoes, Basil is often grown alongside them as companion plants beneficial to growth. A pot of Basil grown indoors will also act an insect repellent.

Virginian Skullcap
Scutellaria lateriflora

Perennial up to 100cm high, with opposite, ovate to lanceolate, toothed leaves. Flowers blue, 2-lipped and borne in 1-sided, axillary spikes. Each calyx has a distinctive, shield-shaped flap on the upper side. Native to North America. Virginian Skullcap is considered to be one of the best sources of nervines or nerve tonics. It contains the sedative scutellarin and is used in small doses to treat epilepsy, insomnia, depression, St Vitus's Dance and other types of nervous conditions. The leaves can be made into a tea but nowadays the herb is usually offered in tablet form.

White Horehound
Marrubium vulgare

White-felted perennial 30–60cm high, with opposite, rounded, wrinkled leaves. Flowers white with a deeply bifid upper lip and a calyx-tube with 10 tiny, hooked teeth at the rim. Native from Europe and North Africa to central Asia. The aerial parts yield bitter principles and a volatile oil, and can be taken as a hot or cold infusion or as a syrup. The plant has been used for heart, liver, and digestive problems and as a quinine substitute for malaria but its main use is for respiratory conditions. Horehound candy is sold as cough sweets.

Balm
Melissa officinalis

Also called Lemon Balm, this 20–70cm high perennial has opposite leaves which smell strongly of lemon when bruised. Flowers 80–150mm long, pale yellow, white or pinkish, 2-lipped. Native to southern Europe, North Africa and western Asia, widely cultivated and introduced elsewhere. As a cordial and tea Balm has a long history, especially as a sedative and for treating nervous conditions and viral infections. The oil is used in aromatherapy and in massage where its antihistamine action is effective against eczema. The fresh leaves are used to flavour drinks, meats, jellies, jams and preserves.

DEAD NETTLES

Motherwort
Leonurus cardiaca

Strong-smelling perennial, 30–200cm, the opposite leaves cut into 3–7 toothed lobes which radiate from the leaf base. Flowers 8–12mm long, white or pale pink with densely hairy upper lips, are carried in compact whorls. Found in most of Europe. The main uses of this herb are for menstrual problems, and postnatal and menopausal anxiety. It is also used to treat rapid or irregular heart beat. The flowering herb makes a bitter tea and it is more normally taken as a syrup or in tablet form. It must not be used during pregnancy.

Betony
Stachys officinalis

Perennial 15–60cm high with sparsely leafy stems but a well-developed basal rosette of long-stalked, oblong leaves. Flower whorls are crowded into spikes. Corollas 15mm long, bright, reddish-purple and 2-lipped, the upper one flat. Native to Europe. An ancient herb, long held in high regard for its supposedly curative and protective powers, Betony has fallen into disuse. Among its many traditional uses is a poultice of fresh leaves to clean wounds, and the dried leaves, which provoke violent sneezing, to clear head colds.

Hyssop
Hyssopus officinalis

Aromatic perennial or miniature shrub 20–60cm high, the stems woody at the base and with narrow, opposite leaves 10–50mm long. Loose whorls of 2-lipped, blue or violet flowers 7–12mm long are carried in slender spikes at the stem tips. Native to southern Europe, North Africa and western Asia, cultivated and introduced elsewhere. The volatile oil found in Hyssop contains the same bitter principle (marrubin) found in White Horehound and medicinally the two plants are used for similar purposes. Only small doses are required and Hyssop should not be used during pregnancy.

Summer Savory
Satureja hortensis

Annual 10–25cm high with narrow opposite leaves 10–30mm long. Flowers white, pink or lilac and carried in rew-flowered whorls. Both calyx and corolla are 2-lipped, the calyx with the lower teeth slightly longer than the upper. Native to the Mediterranean region, but also widely cultivated. A culinary herb with a strong, hot and peppery taste. Commercially the leaves provide a flavouring for salami. Domestically, they are used mainly with vegetables and rich meats.

DEAD NETTLES

Oregano
Origanum vulgare

Often purple-tinged, rather woody and aromatic perennial up to 90cm high with opposite, stalked leaves. Flowers 4–7mm long, white or purplish-pink, 2-lipped. Flowers are carried in small spikes which are themselves crowded into terminal, flat-topped clusters. Native to Europe, northern and western Asia. Oregano is little used medicinally, but the essential oil is used in perfumes, cosmetics and some liqueurs. Used fresh or dried, the leaves of Oregano (also called Marjoram) have a variety of culinary uses, particularly in Italian and Mediterranean cuisines and in various meat products such as sausages.

Sweet Marjoram
Origanum majorana

Similar to, and often confused with, its close relative Oregano, Sweet Marjoram is generally smaller in all its parts, with most leaves more or less stalkless. Flowers have a 1-lipped calyx deeply slit on one side, not with 5 equal teeth as in Oregano. Native to North Africa and south-western Asia, cultivated elsewhere and naturalised in southern Europe. As with Oregano, the essential oil is used in perfumes and cosmetics. More delicately flavoured and sweetly scented than its relative, the fresh or dried leaves are used in lighter dishes with eggs, cream or vegetables.

Thyme
Thymus vulgaris

Miniature aromatic shrub 10–30cm high with erect or spreading branches and narrow, greyish-green, opposite leaves. Flowers white to pale purple, 2-lipped. Native to the western Mediterranean region but cultivated elsewhere. Thyme oil has rather limited medicinal uses, mainly in mouthwashes and cough medicines. Also used in aromatherapy, some cosmetics and toothpastes. It can be toxic when used internally and must not be taken during pregnancy. A widespread and popular culinary flavouring, the fresh or dried leaves can be added to almost any meat or savoury dish.

Pennyroyal
Mentha pulegium

Creeping, often mat-forming perennial 10–40cm high. Opposite leaves 8–30mm, smell like peppermint. Flowers small, lilac and carried in dense whorls in axils of the upper leaves. Found in much of Europe and North Africa. Although rather too strong-smelling for many people, Pennyroyal leaves can be used in stews and stuffings. A homeopathic essence prepared from the fresh herb, or a hot infusion, can be used for coughs and asthmatic problems. However, the essential oil can be toxic and must never be taken during pregnancy.

DEAD NETTLES

Peppermint
Mentha × piperita

Perennial 30–80cm high, varying from almost hairless to grey-woolly. Leaves lanceolate and stalked with a pungent peppermint scent. Flowers lilac-pink and carried in an oblong spike. A hybrid between the wild Water and Spear Mints, it is widely cultivated and naturalised. The leaves are usually used in a fresh state. They yield an essential oil consisting principally of menthol, which has anaesthetic, anti-bacterial and anti-inflammatory properties. It is widely used in inhalants, massage and aromatherapy oils, and to mask the taste of pharmaceutical medicines. Peppermint is also used as a confectionery flavouring.

Red Bergamot
Monarda didyma

Aromatic perennial up to 150cm high with square stems and opposite leaves. Flowers red, tubular and 2-lipped, crowded together in dense terminal clusters. Native to North America, cultivated in Europe. The volatile oil has a scent like that of the Bergamot Orange and the two plants are used in similar ways. The oil is used in aromatherapy to treat anxiety, depression and infections. It may cause uneven pigmentation if used unadultered on the skin. The leaves make the drink Oswego tea, named after the Oswego Indians of North America who first used it.

Lavender
Lavandula angustifolia

Much-branched, aromatic, evergreen shrub up to 100cm with narrow, opposite leaves 2–4cm long. Initially white-hairy, they later turn green. Flowers lavender-blue or purplish, 2-lipped and carried in dense, narrowly cylindrical spikes. Native to the Mediterranean region. Lavender oil is an excellent first-aid remedy for bites, stings and burns. The essential oil content is highest in the flowers. It is used in perfumes and cosmetics, and in aromatherapy to treat infection and stress. Dried flowers placed in lavender bags keep linen fresh. The leaves and flowers are used in herbal teas and tobacco.

Rosemary
Rosmarinus officinalis

Aromatic, evergreen shrub up to 200cm. Opposite leaves 15–40mm long, dark green above and white hairy beneath; narrow and leathery. Flowers 10–12mm long pale blue, 2-lipped, with 2 protruding stamens. Native to the Mediterranean region but cultivated elsewhere. The essential oil distilled from the leaves and flowers is added to pain-relieving liniments and can be applied directly for headaches. It has a reputation as a hair tonic and is a constituent of many shampoos. Rosemary leaves add flavour to a variety of meats, especially lamb.

Sage
Salvia officinalis

Somewhat aromatic, greyish shrub up to 60cm, with woolly branches. Opposite leaves wrinkled above and densely hairy beneath. Flowers up to 35mm long, violet-blue, pink or white and 2-lipped. Native to parts of the east and west Mediterranean but widely cultivated elsewhere. Though less used than formerly, Sage is still regarded as an effective treatment for colds, mouth and throat infections, and to combat the hot flushes of the menopause. It also improves the keeping quality of meat and processed foods and the leaves are added to various sausages, stuffings, pickles, cheeses and even honey.

Deadly Nightshade
Atropa belladonna

Shrubby-looking perennial 50–150cm high. Leaves 60–120mm long, ovate. Flowers violet-brown to greenish, bell-shaped and drooping. Fruits 10–20mm diameter, glossy-black when ripe. A woodland plant native to Europe, North Africa and parts of Asia. Widely used in proprietary medicines. All parts contain the narcotic alkaloid atropine which is used as a sedative and as an antispasmodic for paralysing parts of the nervous system. In eye-drops it dilates the pupils and is used in opthalmology. An extremely poisonous plant which must never be used without medical supervision.

Sweet Pepper
Capsicum annum

Annual 30–90cm high with bright green leaves. Flowers white, drooping, with a loose cone of bluish-yellow stamens. Fruits up to 270mm long, firm and fleshy, ripening from green to yellow or bright red. Native to tropical America, cultivated in most warm and even temperate regions. The fruits of this plant are well-known as a fresh 'vegetable' but when dried and ground the flesh yields paprika. Available in many grades, from mild-to hot-tasting, paprika is used to flavour a wide variety of dishes.

Chili Pepper
Capsicum frutescens

Tall, woody-stemmed perennial up to 200cm. Leaves and flowers similar to those of the closely related Sweet Pepper but fruits are generally smaller, narrower, sometimes twisted, and may be yellow, orange or red. Native to South America but widely cultivated throughout the tropics. Cayenne is the dried and ground fruits of the Chili Pepper. It is very warming and is used externally to treat muscle and nerve pain, and internally to stimulate the circulation and treat colds. It has many culinary uses and is a principal ingredient in Chili powder and Tabasco sauce.

NIGHTSHADES

Thornapple
Datura stramonium

Annual 50–200cm high, with coarse, wavy-toothed leaves. Flowers 70–120mm, white to pale-violet, fragrant and trumpet-shaped. Seeds contained in spiny, 4-chambered capsules. Native to North America but naturalised in many parts of the world. Thornapple is related to both Deadly Nightshade and Henbane. It contains similar alkaloids, and is likewise a narcotic. It is also a painkiller and has been used as an anaesthetic and to treat Parkinson's disease. Despite its medicinal uses, all parts of the plant are highly poisonous and must never be eaten.

Mandrake
Mandragora officinalis

Rosette-forming perennial up to 15cm high with a deep, forked taproot and dark green leaves up to 300mm long. Flowers greenish-white and bell-shaped. Fruits globose resembling tomatoes, ripening from green to yellow. Native to central and south-east Europe, rare in the wild but cultivated in places. With its unusual man-shaped root, Mandrake was one of the magical plants of the ancient herbalists. It is, however, a genuine medicinal herb, the root yielding a strong anaesthetic still used in modern medicine as a preoperative drug. Like its relative Deadly Nightshade, it is poisonous.

Sesame
Sesamum orientale

Erect annual up to 60cm, with white, pink or mauve, trumpet-shaped flowers. Fruit is a 30mm-long capsule containing numerous shiny, ovoid seeds. Native to tropical Asia but widely cultivated in other hot areas. Sesame seeds, high in protein and oil, have a nutty flavour when cooked. They are used as a garnish or ground and added to various dishes, particularly in eastern Mediterranean cuisines. The seeds are often sprinkled on bread or biscuits. The polyunsaturated oil expressed from the seeds is used for cooking and makes a margarine suitable for low-cholesterol diets.

Great Mullein
Verbascum thapsus

Tall biennial up to 200cm, the whole plant densely covered with white or greyish down. Flowers 12–35mm across, yellow, crowded in a long terminal spike. Upper stamens have white hairs on the filaments. Native to Europe and Asia, naturalised in North America. The leaves provide a tea which, when strained, soothes coughs and bronchial complaints. Fresh leaves make an effective compress for wounds, burns or chilblains while dried leaves are added to herbal tobacco and smoked to ease asthma. The flowers softened in olive oil can be used for ear-drops or as a liniment.

PLANTAINS

Foxglove
Digitalis purpurea

Biennial or perennial up to 180cm, with softly hairy basal leaves. Bell-shaped flowers 40–55m long form a long spike. Petals purple, pink or white and usually black-spotted on the inside. Native to western Europe. The leaves of Foxglove yield the drug digitalin which contains several compounds which affect the cardiac muscle to increase the heartbeat. Extremely toxic, it can cause paralysis and sudden death if misused. Foxglove was not properly recognised as a medicinal herb until the late eighteenth century. The similar Woolly Foxglove (D. *lanata*) from southern Europe has now largely replaced it in commercial production.

Greater Plaintain
Plantago major

Perennial up to 60cm, the large, strongly veined leaves forming a basal rosette. Flowers small, greenish-yellow, form a terminal spike on a stalk at least as long as the leaves. Native to Europe, North Africa and parts of Asia but now spread through most temperate regions of the world. The bruised or crushed leaves are styptic, helping to staunch bleeding, ana will draw the pain from bites, stings or burns. They contain mucilage, tannins and silica, and an infusion is used to treat bronchitis, coughs and lung problems.

Eyebright
Euphrasia rostkoviana

Erect, branched semi-parasitic annual up to 35cm, with opposite, toothed leaves. Flowers 8–12.5mm, upper lip often lilac, lower lip white with yellow markings. Found throughout most of Europe and a few adjacent areas. There are many closely related, similar species. As the name suggests, the aerial parts of this herb are used to make a soothing eyewash, and it is also effective against hayfever, colds and catarrh. An infusion can be used externally or, diluted, taken internally for the same symptoms.

Pepper
Piper nigrum

Woody climber up to 600cm high with smooth, twining stems and large, thick and leathery leaves. Flowers small, greenish and petal-less, carried in long drooping spikes. Berries ripen from green to orange then red. Native to tropical Asia, but widely cultivated in the tropics. An essential condiment, pepper is one of the earliest and most valued of the eastern spices. Black peppers are the dried, unripe berries. Soaking and removing the outer skin of the unripe berries yields the milder white peppers.

Elder
Sambucus nigra

Small, bushy tree or shrub with fragrant white flowers but foetid, unpleasant-smelling foliage. Opposite leaves pinnate, with 5–7 leaflets. Flower heads 10–24cm across, branched and flat-topped, nodding when the black berries are ripe. Native to most of Europe, North Africa and western Asia. Elder flowers are used as an infusion to treat catarrh and hayfever, as an eyewash and gargle, and to make a skin ointment. Both flowers and berries are traditionally used to make wines and cordials, and the berries are excellent in jams and pies. All other parts of the plant are poisonous.

Valerian
Valeriana officinalis

Downy perennial up to 200cm, usually unbranched, with pinnate or pinnately lobed leaves. Flowers pale pink and funnel-shaped, unequally 5-lobed and carried in a compound head made up of smaller, dense heads. Found in woods and grassy places from Europe to Japan. The roots are dried then macerated in cold water. Another of the sedative drugs, Valerian acts on the central nervous system and is good for anxiety, tension and nervous headaches. It also lowers blood pressure. If used over long periods the drug is reported to become addictive.

Boneset
Eupatorium perfoliatum

Perennial 30–120cm high. The bases of each pair of lanceolate, wrinkled leaves are fused to encircle the stem. Flower heads with a few white or pale purple flowers are carried in flat clusters. Native to North America. Leaf tea was traditionally and widely used by Indians and early settlers to treat fevers and influenza. It was claimed to be particularly successful in the United States during the influenza epidemics of the 19th Century. Modern research suggests that the herb stimulates the immune system. It can be toxic in large doses.

Goldenrod
Solidago virgaurea

Downy perennial up to 100cm with lanceolate to ovate leaves widest above the middle. Flower heads yellow, with both disc and ray florets, and are carried in branched spikes. Found in a variety of habitats throughout the temperate northern hemisphere. A mild diuretic, this herb is used in homeopathy and in numerous proprietary medicines for kidney and bladder disorders, as well as for arthritis and rheumatism. It can be taken as an infusion made from the aerial parts collected before the flowers fully develop.

DAISIES

Yarrow
Achillea millefolium

Erect perennial up to 60cm, downy, aromatic and with much-divided, dark green leaves. Flower heads up to 10mm across in flat-topped clusters. A plant of grassy places, native to Europe and western Asia, introduced elsewhere. A traditional herb used throughout the northern hemisphere. Tea made from the dried flowering plant is used to treat colds and fevers. The plant has styptic properties, helping to control both internal and external bleeding and helps clear blood clots. It contains at least one toxic compound and in large doses may cause photosensitive skin reactions.

Elecampane
Inula helenium

Tall, downy perennial up to 250cm, the lower leaves stalked, the upper stalkless and clasping the stem. Flower heads up to 60–80mm across, florets yellow. Native to southeastern Europe and western Asia, cultivated and widely naturalised in other temperate regions. Tea made from the roots is a traditional treatment for asthma, bronchitis, pneumonia and whooping cough and as a wash counters neuralgia, sciatica and skin diseases. It was formerly used to treat tuberculosis. In Chinese medicine the flowers are used to treat some forms of cancer. Elecampane is also used in confectionery and to flavour drinks.

Sunflower
Helianthus annuus

Tall, stout annual reaching 300cm, sometimes more. Flower heads may reach 300mm across with golden ray florets surrounding the brown disc. Seeds often striped black and white. Native to North America but cultivated everywhere both on a commercial scale and as a garden ornamental. All parts are used, for various purposes. Tea made from the flowers treats lung problems and malaria, that made from the leaves treats fever and bites; both can cause allergic reactions in some people. Sunflowers are best known, though, as a source of high quality, edible oil, obtained by crushing the seeds.

Scented Mayweed
Matricaria recutita

Also called German Chamomile.

A strongly aromatic annual up to 60cm, with much-divided leaves. Flower heads 30–50mm across, the white ray florets turned downwards. The yellow central disc is high-domed and hollow. Probably native to southern and eastern Europe and parts of Asia but a widespread weed. This herb is a constituent of skin ointments and shampoos and has an unsubstantiated reputation as a treatment for cancer. Tea made from the dried flowers is used for insomnia, hyperactivity and anxiety. Individuals who are allergic to Ragweeds may suffer a similar reaction to this herb.

Chamomile
Chamaemelum nobile

Also called Roman Chamomile.

A hairy, aromatic perennial 10–30cm. Similar in appearance to Scented Mayweed out flower heads 18–25cm across have a conical, solid disc of yellow florets and the white, outer ray florets are sometimes lacking. Native to western Europe and North Africa but often cultivated and naturalised elsewhere. It contains similar compounds to, and is sometimes used in place of, Scented Mayweed: the blue oil extracted from Chamomile is both more acrid and less effective than that of its close relative. Because it forms mats, it is the species planted for Chamomile lawns.

Tansy
Tanacetum vulgare

Strongly aromatic perennial up to 150cm high with glandular, pinnately lobed leaves. Flower heads yellow, button-like with up to 100 in each flat-topped cluster. Found throughout most of Europe and much of northern Asia out often as an escape from cultivation. The dried aerial parts are a traditional insecticidal and vermifuge herb, formerly used both internally and externally. Except in homeopathy, internal use of Tansy is now discouraged as it is poisonous, the essential oil especially proving fatal if taken even in small doses. In the United States it is illegal to sell this herb.

Feverfew
Tanacetum parthenium

Yellowish-green, aromatic perennial up to 60cm with pinnately lobed leaves. Flower heads 10–25mm across. Native to the Balkan Peninsula and western Asia but long cultivated for medicine and naturalised in many regions of the world. A sedative tea made from the leafy parts is a traditional treatment for arthritis, colds and cramp but recently the plant has received much attention as a cure for migraine. However it should be used with care as it can cause an allergic reaction, dermatitis and, particularly, mouth sores.

Costmary
Balsamita major

Dull green, densely hairy perennial up to 120cm high with large oblong, finely toothed leaves. Flower heads 10–16mm across carried in branched clusters. The white ray florets are sometimes absent, giving the heads a button-like appearance. Native to western Asia but widely introduced in Europe and North America. Originally a brewing herb used to flavour beers; the leaves and flowers of Costmary are still sometimes used to flavour meat and cakes, or to make a tonic tea. It has a minty or balsam-like scent and strong flavour, so little is required.

DAISIES

Wormwood
Artemisia absinthium

Aromatic, woody perennial 30–90cm high. Leaves white, silky-hairy, 2–3 pinnately lobed. Flower heads 3–5mm across containing only disc florets. Found in most of Europe though probably only introduced in some areas, as it is in North America. Wormwood is a vermifuge which is nowadays mainly used as a tonic and digestive aid. It formerly provided the bitter flavouring in the liqueur Absinthe, but it is now banned in alcoholic drinks as it contains the compound thujone which irreparably damages the central nervous system. With the thujone removed, however, Wormwood is an approved food flavouring.

Southernwood
Artemisia abrotanum

Strongly aromatic shrub up to 100cm with finely divided, gland-dotted leaves grey-hairy beneath. Flower heads 3–4mm across, yellowish and button-like. Of uncertain origin, it is widely cultivated and naturalised, especially in southern Europe. Closely related to Wormwood, Southernwood is similarly used as a tonic, to expel worms and, when dried, to make a moth repellant. The tonic aids menstruation and should not be used during pregnancy. The young leaves and shoots, bitter but lemon-tasting with a strong lemon scent, are used for flavouring cakes.

Tarragon
Artemisia dracunculus

Aromatic, hairless perennial 60–120cm high, narrow leaves mostly entire but the lower ones 3-toothed at their tips. Flower heads small, yellow, globose, drooping on downcurved stalks. Native to southern and eastern parts of Russia, widely cultivated and naturalised in many other areas. Unlike the related Wormwood and Southernwood, Tarragon is used solely as a culinary herb. Introduced into Britain in the mid-fifteenth century. Tarragon is used in many sauces, *fines herbes* mixes, marinades and preserves, particularly in French cuisine. It is best used fresh, or with the flavour preserved in oil or vinegar.

Colt's foot
Tussilago farfara

Perennial producing flowering stems up to 15cm high before the large, round, shallowly lobed and toothed leaves appear. Leaves up to 300mm when mature, green above, white woolly beneath. Found in damp waste places in most of the temperate northern hemisphere. A soothing tea or syrup made from the leaves and flowers acts on the mucus membranes and is a widespread cure for coughs and bronchial problems. The dried leaves can be smoked to produce a similar effect and are thought to have an antihistamine action. The plant is potentially toxic in large doses.

DAISIES

Marigold
Calendula officinalis

Much-branched perennial up to 70cm with oblong to spoon-shaped, glandular downy or sparsely woolly leaves. Flower heads 40–70mm across, yellow or orange. Possibly originating in central and southern Europe but cultivated in many regions. The flowers are antiseptic and as a compress can be used for burns and ulcers. The edible petals, collected as the flowers are opening, are added to salads and stews. They can be used as a substitute for Saffron in colouring cheese, butter and cakes.

Burdock
Arctium lappa

Coarse, downy biennial reaching 150cm. Leaves broad with heart-shaped bases and solid stalks. Flower heads 30–40mm across, prickly, containing only purple disc florets. A plant of shady places in Europe and parts of western Asia. Burdock is said to purify the blood, and is also used to treat boils, measles and other skin eruptions. It is usually taken as a tea prepared from just the roots or the roots and leaves together. The seeds are used for abscesses, bites and bruises. Burdock is still used to make Dandelion and Burdock beer.

Blessed Thistle
Cnicus benedictus

Cobweb-hairy annual 10–60cm high with pale green, pinnately lobed leaves up to 300mm long, the lobes pointing backwards and fringed with small, spiny teeth. Stem leaves are spine-tipped. Flower heads large with yellow florets and surrounded by the upper leaves. A Mediterranean weed, cultivated and naturalised elsewhere. Although slightly toxic in large doses, causing nausea, a weak tea made from the flowering plant is said to be beneficial for a wide variety of medical problems, from colds and migraines to jaundice, ringworm and even deafness. Mainly used as a digestive tonic and appetite reviver.

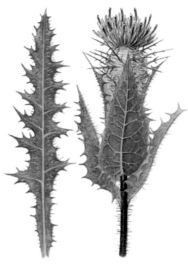

Safflower
Carthamus tindorius

Spiny, thistle-like annual up to 60cm high. Basal leaves pinnately lobed but stem leaves are undivided. Flower heads have spiny, leaf-like involucral bracts and numerous yellow, orange or red florets. Native to western Asia but cultivated and often naturalised in southern and central Europe. Safflower tea was used to treat measles but it is principally a dye-plant. The flowers yield two colours, red for dyeing silk and yellow for colouring food. It is sometimes used as a Saffron substitute. Nowadays the seeds provide a source of dietary oil which is low in cholesterol.

DAISIES

Chicory
Cichorium intybus

Blue-flowered perennial with branched stems up to 120cm high. Basal leaves pinnately lobed, stem leaves may be entire and clasp the stem. Flower-heads, containing ray florets only, are 25–40mm across and carried in small groups. Native to Europe, North Africa and western Asia, introduced into most other temperate regions. Extracts from Chicory roots have been used medicinally, mainly as diuretics and laxatives, and have a depressive effect on heart rate. The root is best known used as an additive to or even substitute for coffee. The very young, blanched leaves are a popular salad vegetable.

Dandelion
Taraxacum officinale

Variable perennial up to 50cm high, with a rosette of usually pinnately lobed leaves. Flower-heads 350–500mm, on stout, hollow stalks; only yellow ray florets are present. A widespread plant of temperate grassland and waste ground, often a weed. The leaves and fresh root make a powerfully diuretic tea while the weaker dried root can be roasted to make a coffee substitute. The whole plant is edible, the leaves and flowers rich in vitamins, especially A and C. The flowers are used for wine and the buds can be pickled like capers.

Aloe
Aloe vera

Stemless perennial with creeping
stolons, forming clusters of leaf-rosettes.
Leaves 35–60cm, blue-green and spiny,
sometimes tinged with red. Flowers
25–30mm long, drooping and cylindrical,
are carried in spikes up to 50cm. A coastal
plant native to the Mediterranean region.
Often encountered as an ingredient of
shampoos and cosmetics, Aloe yields
a soothing medicinal gel extracted
from the leaves. This gel stimulates
regeneration of the skin when smeared
onto cuts or burns, and Aloe has the
alternative common name of First-aid-
plant. The juice from the cut eaves is a
very strong emetic.

Lily-of-the-Valley
Convallaria majalis

Perennial up to 37cm with ovate leaves
arising directly from a creeping rhizome,
their sheathing bases forming the
stem. Flowers white or pink, fragrant
and carried in an erect, 1-sided spike.
Native to Europe and north-east Asia
but a common garden plant. The leaves
of Lily-of-the-Valley contain cardiac
glycosides similar to those of Foxglove,
and which have a similar effect in
strengthening and regulating the
heartbeat. It is an alternative treatment
for cardiac patients also suffering from
high blood-pressure. All parts of the
plant are poisonous and it must be used
only under medical supervision.

ONIONS

Chives
Allium shoenoprasum

Tuft-forming perennial up to 50cm high
with narrowly conical bulbs less than
1cm in diameter attached to a short
rhizome. Stem and 1–2 slender leaves
are cylindrical and hollow. Flowers
7–15mm long, lilac to pale purple or
rarely white, bell-shaped and crowded
in a dense umbel. Found in most of
the northern hemisphere and widely
cultivated. Purely a culinary herb,
Chives have a milder and much more
delicate flavour than the related onions.
Unlike onions and Garlic, it is the finely
chopped leaves that are used, not the
small, thin bulbs.

Garlic
Allium sativum

Perennial with stems up to 100cm high.
Bulbs 3–6cm in diameter and composed
of 5–15 small bulbs enclosed in a papery
sheath. Leaves up to 60cm long, flat
and grass-like. The umbel contains a
few cup-shaped, white, pink or purple
flowers but many bulbils. Native to
central Asia but widely cultivated.
It is also an ancient medicinal herb,
reducing blood pressure, clots, sugar-
and cholesterol-levels. Its antiseptic and
antibiotic properties are employed in
treating infected wounds and intestinal
disorders. Best known as a culinary herb
it is used in many dishes.

Saffron
Crocus sativus

Autumn-flowering species, the grass-like leaves appearing after the flowers. Flowers lilac-purple with a yellowish throat; goblet-shaped. The prominent 3-branchea style is orange. Only the large styles are used; at least 60,000 flowers are required to yield 1 pound of the spice, and it has always been very expensive. A well-known food dye, Saffron imparts a sweet scent as well as an orange-yellow colour to rice dishes, soups and cakes. Nowadays, inferior but cheaper substitutes such as Turmeric and Safflower are often used instead.

Sweet Flag
Acorus calamus

Perennial up to 125cm, the linear leaves with wavy margins and aromatic when crushed. Flowers yellowish-green, tiny and packed into a compact, up-curved spike. Native to southern and eastern Asia but naturalised in Europe and North America. The rhizome has various medicinal uses. Chewed raw or made into a tea, it reduces stomach acidity and, because it stimulates digestive secretions and therefore appetite, is recommended for anorexia nervosa. It is also used to help smokers to give up tobacco. Some strains contain a carcinogen but in others the chemical is absent.

TRUE GRASSES

Couch Grass
Elytrigia repens
Dull-to bluish-green perennial with tough, far-creeping rhizomes. Flowering stems up to 120cm high, hairless and stiffly erect. Flowering spikes 50–200mm long, slender, unbranched and composed of paired spikelets. Found throughout much of the northern hemisphere, often as a pernicious weed. The pale rhizomes are made into a tea or decoction. A diuretic rich in minerals and vitamins A and B, it has antibiotic properties and is used to treat incontinence, kidney stones and other problems of the urinary tract. In Africa it is regarded as an antidote to arrow poisons.

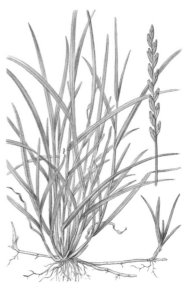

Lemon Grass
Cymbopogon citratus
Clump-forming, aromatic perennial with flowering stems reaching 2m. Leaves up to 60cm × 0.5cm are tapered at both ends. Flowers carried in a large plume-like panicle with a drooping tip. A native of southern India and Sri Lanka but widely cultivated in the tropics and occasionally elsewhere. The scent is due to the presence of citral, and the oil distilled from Lemon Grass is used in the perfume industry and the artificial synthesis of vitamin A. The strongly scented leaves are increasingly popular in culinary use for imparting a lemon flavour to dishes.

Ginger
Zingiber officinale

Perennial up to 100cm with short, fleshy and knobbly, branched rhizomes. Stems erect with 2 ranks of leaves, the sheathing leaf-stalks forming the stem itself. Flowers, yellow or white with a purple lower lip, form dense spikes. Native to tropical south-east Asia but also cultivated in Africa and the Caribbean. It makes a warming and mildly stimulating treatment for the circulation and for bronchial complaints. The rhizome, either fresh or dried, peeled or not, is a culinary spice; the crystallised stem is also eaten.

Turmeric
Curcuma domestica

Perennial up to 100cm similar to its close relative Ginger but the leaves are all basal, their sheathing stalks not, or scarcely, forming a stem. Rhizomes *c.* 25mm diameter, yellowish on the outside and deep orange within. Probably native to India, cultivated there and in other parts of the tropics. The boiled, dried and powdered rhizome has a characteristic pungent smell. Widely used as both a flavouring and as a colouring agent for imparting a brilliant yellow to a variety of dishes, it is an essential ingredient in curries.

Cardamom
Elletaria cardamomum

Perennial up to 350cm. Related to Ginger, it also has thick, fleshy rhizomes and very tall sterile stems formed by the sheathing leaf-stalks. Flowering stems leafless and spreading. Flowers white with blue and yellow markings and a single protruding stamen. Capsules 10–20mm long, ovoid, greenish-grey, are harvested unripe and dried whole. Each contains 3–4 brown seeds. Native to hills of southern India but also cultivated in Sri Lanka and parts of Central America. Dried, ground seeds are used to flavour sweet and savoury dishes and for spicing wine but are best known as an ingredient in curries.

Vanilla
Vanilla planifolia

Evergreen, climbing orchid, up to 30m, with fleshy leaves. Flowers 40–70mm long, greenish-yellow with an inrolled, orange-striped lower lip. Pods up to 200mm long are fragrant when ripe and contain numerous tiny seeds. Native to Central America but cultivated throughout the tropics, principally in Madagascar. Outside its native range, the flowers must be hand-pollinated. First used by the Aztecs to flavour chocolate drinks. The unique flavour is due to vanillin crystals on the surface of the pods. Pods are still used for culinary flavouring, despite the availability of synthetic essence.

Image gallery

▲ Great Yellow Gentian, p.62

▲ Surinam Quassia-wood, p.47   ▼ Marsh-mallow, p.49

▲ Sweet Violet, p.50   ▼ Witch Hazel, p.52

▲ Fennel, p.56　　　　　　　　　▼ Angelica, p.59

▲ Borage, p.66

▼ Motherwort, p.70

▲ Chili Pepper, p.77   ▼ Elder, p.82

▲ Safflower, p.91

BIBLIOGRAPHY

Blackwell, W.H., *Poisonous and Medicinal Plants*. Prentice Hall, New Jersey, 1990.

British Pharmacopoeia, vol I–II. HMSO, London, 1980.

Duke, J.A. and Ayensu, E.S., *Medicinal Plants of China*, vol I–II. Reference Publications, Algonac, Michigan, 1985.

Foster, S. and Duke, A., A *Field Guide to Medicinal Plants: Eastern and Central North America*. Houghton Mifflin, Boston, 1990.

Jain, S.K. and DeFilipps, R.A., *Medicinal Plants of India*, vol I–II. Reference Publications, Agonac, Michigan, 1991.

Launert, E., *Edible and Medicinal Plants of Britain and Northern Europe*. Hamlyn, London, 1981.

Lou, Z., el al (eds), *Colour Atlas of Chinese Traditional Drugs*, vol I. Science Press, Beijing, 1987.

Mabey, R., *The New Age Herbalist*. Collier Books, New York, 1988.

Phillips, R. and Foy, N., *Herbs*. Pan, London, 1990.

Stace, C.A., *New Flora of the British Isles*. Cambridge University Press, Cambridge, 1991.

Stuart, M., (ed.) *The Encyclopedia of Herbs and Herbalism*, Orbis, London, 1985.

Tutin, T.G. el al (eds), *Flora Europea*, vols I–V. Cambridge

ORGANISATIONS

American Herb Society, www.herbsociety.org

The Australian Herb Society Incorporated, www.herbsociety.org.au

Herb Federation of New Zealand, www.herbs.org.nz

The Herb Society, www.herbsociety.org.uk

The National Herbalists Association of Australia, www.nhaa.org.au

The National Institute of Medical Herbalists, www.nimh.org.uk

HERB GARDENS AND COLLECTIONS TO VISIT

The Butser Ancient Farm Research Project, www.butserancientfarm.co.uk

Cambridge Botanic Garden, www.botanic.cam.ac.uk

Chelsea Physic Garden, www.chelseaphysicgarden.co.uk

Hatfield House, www.hatfield-house.co.uk

Michelham Priory Physic Garden, www.sussexpast.co.uk/properties-to-discover/ michelham-priory

The Museum of Garden History, www.gardenmuseum.org.uk

Ness Botanic Gardens, www.nessgardens.org.uk'

The Royal Botanic Gardens, www.kew.org

Royal Botanic Garden, www.rbge.org.uk

The Royal Horticultural Society Gardens, www.rhs.org.uk

The Tudor Garden Museum, www.tudorhouseandgarden.com

Index

Page numbers in **bold** refer to photographs

Index

Index